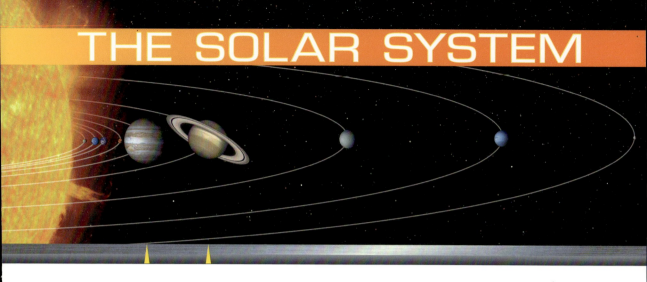

THE SOLAR SYSTEM

JUPITER AND SATURN

REVISED EDITION

Linda T. Elkins-Tanton

Facts On File
An imprint of Infobase Publishing

*To my mother and father, Sally and Leonard Elkins, and
to my brother and sister-in-law, James P. Elkins and
Margaret MacNamidhe Elkins*

JUPITER AND SATURN, Revised Edition

Copyright © 2011, 2006 by Linda T. Elkins-Tanton

Facts On File, Inc.
An imprint of Infobase Publishing
132 West 31st Street
New York NY 10001

Library of Congress Cataloging-in-Publication Data
Elkins-Tanton, Linda T.
    Jupiter and Saturn / Linda T. Elkins-Tanton ; foreword, Maria T. Zuber.—Rev. ed.
        p. cm.
    Includes bibliographical references and index.
    ISBN 978-0-8160-7698-7
    1. Jupiter (Planet)—Popular works. 2. Saturn (Planet)—Popular works. I. Title.
    QB661.E45 2010
    523.45—dc22   2009045019

Facts On File books are available at special discounts when purchased in bulk quantities for businesses, associations, institutions, or sales promotions. Please call our Special Sales Department in New York at (212) 967-8800 or (800) 322-8755.

You can find Facts On File on the World Wide Web at http://www.factsonfile.com

Excerpts included herewith have been reprinted by permission of the copyright holders; the author has made every effort to contact copyright holders. The publishers will be glad to rectify, in future editions, any errors or omissions brought to their notice.

Text design by Annie O'Donnell
Composition by Hermitage Publishing Services
Illustrations by Dale Williams and Pat Meschino
Photo research by Elizabeth H. Oakes
Cover printed by Bang Printing, Brainerd, Minn.
Book printed and bound by Bang Printing, Brainerd, Minn.
Date printed: October 2010
Printed in the United States of America

10 9 8 7 6 5 4 3 2 1

This book is printed on acid-free paper.

Contents

Foreword

While I was growing up, I got my thrills from simple things—one was the beauty of nature. I spent hours looking at mountains, the sky, lakes, et cetera, and always seeing something different. Another pleasure came from figuring out how things work and why things *are* the way they *are*. I remember constantly looking up things from why airplanes fly to why it rains to why there are seasons. Finally was the thrill of discovery. The excitement of finding or learning about something new—like when I found the Andromeda galaxy for the first time in a telescope—was a feeling that could not be beat.

Linda Elkins-Tanton's multivolume set of books about the solar system captures all of these attributes. Far beyond a laundry list of facts about the planets, the Solar System is a set that provides elegant descriptions of natural objects that celebrate their beauty, explains with extraordinary clarity the diverse processes that shaped them, and deftly conveys the thrill of space exploration. Most people, at one time or another, have come across astronomical images and marveled at complex and remarkable features that seemingly defy explanation. But as the philosopher Aristotle recognized, "Nature does nothing uselessly," and each discovery represents an opportunity to expand human understanding of natural worlds. To great effect, these books often read like a detective story, in which the 4.5-billion year history of the solar system is reconstructed by integrating simple concepts of chemistry, physics, geology, meteorology, oceanography, and even biology with computer simulations, laboratory analyses, and the data from the myriad of space missions.

Starting at the beginning, you will learn why it is pretty well understood that the solar system started as a vast, tenuous ball of gas and dust that flattened to a disk with most of the mass—the future Sun—at the center. Much less certain is the transition from a dusty disk to the configuration with the planets, moons, asteroids, and comets that we see today. An ironic contrast is the extraordinary detail in which we understand some phenomena, like how rapidly the planets formed, and how depressingly uncertain we are about others, like how bright the early Sun was.

Once the planets were in place, the story diverges into a multitude of fascinating subplots. The oldest planetary surfaces preserve the record of their violent bombardment history. Once dismissed as improbable events, we now know that the importance of planetary impacts cannot be overstated. One of the largest of these collisions, by a Mars-sized body into the Earth, was probably responsible for the formation of the Earth's Moon, and others may have contributed to extinction of species on Earth. The author masterfully explains in unifying context the many other planetary processes, such as volcanism, faulting, the release of water and other volatile elements from the interiors of the planets to form atmospheres and oceans, and the mixing of gases in the giant planets to drive their dynamic cloud patterns.

Of equal interest is the process of discovery that brought our understanding of the solar system to where it is today. While robotic explorers justifiably make headlines, much of our current knowledge has come from individuals who spent seemingly endless hours in the cold and dark observing the night skies or in labs performing painstakingly careful analyses on miniscule grains from space. Here, these stories of perseverance and skill receive the attention they so richly deserve.

Some of the most enjoyable aspects of these books are the numerous occasions in which simple but confounding questions are explained in such a straightforward manner that you literally feel like you knew it all along. How do you know what is inside a planetary body if you cannot see there? What makes solar system objects spherical as opposed to irregular in shape? What causes the complex, changing patterns at the top of Jupiter's atmosphere? How do we know what Saturn's rings are made of?

When it comes right down to it, all of us are inherently explorers. The urge to understand our place on Earth and the extraordinary worlds beyond is an attribute that makes us uniquely human. The discoveries so lucidly explained in these volumes are perhaps most remarkable in the sense that they represent only the tip of the iceberg of what yet remains to be discovered.

—Maria T. Zuber, Ph.D.
E. A. Griswold Professor of Geophysics
Head of the Department of Earth,
Atmospheric and Planetary Sciences
Massachusetts Institute of Technology
Cambridge, Massachusetts

Preface

On August 24, 2006, the International Astronomical Union (IAU) changed the face of the solar system by dictating that Pluto is no longer a planet. Though this announcement raised a small uproar in the public, it heralded a new era of how scientists perceive the universe. Our understanding of the solar system has changed so fundamentally that the original definition of *planet* requires profound revisions.

While it seems logical to determine the ranking of celestial bodies by size (planets largest, then moons, and finally asteroids), in reality that has little to do with the process. For example, Saturn's moon Titan is larger than the planet Mercury, and Charon, Pluto's moon, is almost as big as Pluto itself. Instead, scientists have created specific criteria to determine how an object is classed. However, as telescopes increase their range and computers process images with greater clarity, new information continually challenges the current understanding of the solar system.

As more distant bodies are discovered, better theories for their quantity and mass, their origins, and their relation to the rest of the solar system have been propounded. In 2005, a body bigger than Pluto was found and precipitated the argument: Was it the 10th planet or was, in fact, Pluto not even a planet itself? Because we have come to know that Pluto and its moon, Charon, orbit in a vast cloud of like objects, calling it a planet no longer made sense. And so, a new class of objects was born: the dwarf planets.

Every day, new data streams back to Earth from satellites and space missions. Early in 2004, scientists proved that standing liquid water once existed on Mars, just a month after a mission visited a comet and discovered that the material in its nucleus is as strong as some *rocks* and not the loose pile of ice and dust expected. The MESSENGER mission to Mercury, launched in 2004, has thus far completed three flybys and will enter Mercury orbit at 2011. The mission has already proven that Mercury's core is still molten, raising fundamental questions about processes of planetary evolution, and it has sent back to Earth intriguing information about the composition of Mercury's crust. Now the New Horizons mission is on its way to make the first visit to Pluto and the Kuiper belt. Information arrives from space observations and Earth-based experiments, and scientists attempt to explain what they see, producing a stream of new hypotheses about the formation and evolution of the solar system and all its parts.

The graph below shows the number of moons each planet has; large planets have more than small planets, and every year scientists discover new bodies orbiting the gas giant planets. Many bodies of substantial size orbit in the asteroid belt, or the Kuiper belt, and many sizable asteroids cross the orbits of planets as they make their way around the Sun. Some planets' moons are unstable and will in the near future (geologically speaking) make new ring systems as they crash into their hosts. Many moons, like Neptune's giant Triton, orbit their planets backward (clockwise when viewed from the North Pole, the opposite way that the plan-

The mass of the planet appears to control the number of moons it has; the large outer planets have more moons than the smaller inner planets.

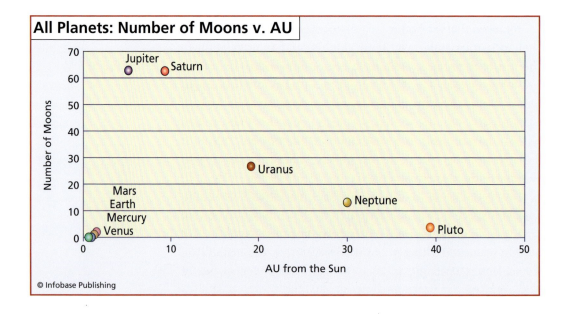

All Planets: Number of Moons v. AU

Number of Moons (y-axis), AU from the Sun (x-axis)

Jupiter, Saturn, Uranus, Neptune, Pluto, Mars, Earth, Mercury, Venus

© Infobase Publishing

ets orbit the Sun). Triton also has the coldest surface temperature of any moon or planet, including Pluto, which is much farther from the Sun. The solar system is made of bodies in a continuum of sizes and ages, and every rule of thumb has an exception.

Perhaps more important, the solar system is not a static place. It continues to evolve—note the drastic climate changes we are experiencing on Earth as just one example—and our ability to observe it continues to evolve, as well. Just five planets visible to the naked eye were known to ancient peoples: Mercury, Venus, Mars, Jupiter, and Saturn. The Romans gave these planets the names they are still known by today. Mercury was named after their god Mercury, the fleet-footed messenger of the gods, because the planet Mercury seems especially swift when viewed from Earth. Venus was named for the beautiful goddess Venus, brighter than anything in the sky except the Sun and Moon. The planet Mars appears red even from Earth and so was named after Mars, the god of war. Jupiter is named for the king of the gods, the biggest and most powerful of all, and Saturn was named for Jupiter's father. The ancient Chinese and the ancient Jews recognized the planets as well, and the Maya (250–900 C.E., Mexico and environs) and Aztec (~1100–1700 C.E., Mexico and environs) knew Venus by the name Quetzalcoatl, after their god of good and light, who eventually also became their god of war.

Science is often driven forward by the development of new technology, allowing researchers to make measurements that were previously impossible. The dawn of the new age in astronomy and study of the solar system occurred in 1608, when Hans Lippenshey, a Dutch eyeglass-maker, attached a lens to each end of a hollow tube and thus created the first telescope. Galileo Galilei, born in Pisa, Italy, in 1564, made his first telescope in 1609 from Lippenshey's model. Galileo soon discovered that Venus has phases like the Moon does and that Saturn appeared to have "handles." These were the edges of Saturn's rings, though the telescope was not strong enough to resolve the rings correctly. In 1610, Galileo discovered four of Jupiter's moons, which are still called the Galilean satellites. These four moons were the proof that not every heavenly body orbited the Earth as Ptolemy, a Greek philosopher, had

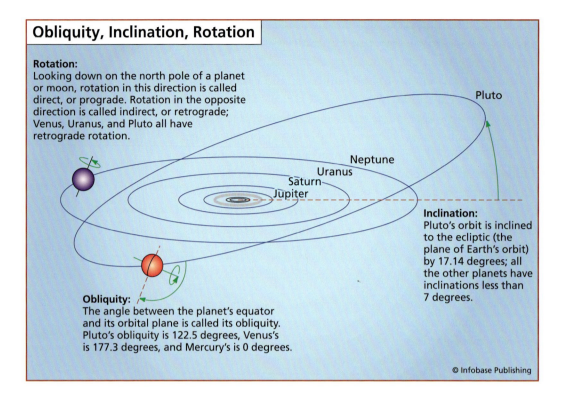

Obliquity, Inclination, Rotation

Rotation:
Looking down on the north pole of a planet or moon, rotation in this direction is called direct, or prograde. Rotation in the opposite direction is called indirect, or retrograde; Venus, Uranus, and Pluto all have retrograde rotation.

Pluto

Neptune
Uranus
Saturn
Jupiter

Inclination:
Pluto's orbit is inclined to the ecliptic (the plane of Earth's orbit) by 17.14 degrees; all the other planets have inclinations less than 7 degrees.

Obliquity:
The angle between the planet's equator and its orbital plane is called its obliquity. Pluto's obliquity is 122.5 degrees, Venus's is 177.3 degrees, and Mercury's is 0 degrees.

© Infobase Publishing

Obliquity, orbital inclination, and rotational direction are three physical measurements used to describe a rotating, orbiting body.

asserted around 140 C.E. Galileo's discovery was the beginning of the end of the strongly held belief that the Earth is the center of the solar system, as well as a beautiful example of a case where improved technology drove science forward.

The concept of the Earth-centered solar system is long gone, as is the notion that the heavenly spheres are unchanging and perfect. Looking down on the solar system from above the Sun's north pole, the planets orbiting the Sun can be seen to be orbiting counterclockwise, in the manner of the original *protoplanetary disk* of material from which they formed. (This is called *prograde* rotation.) This simple statement, though, is almost the end of generalities about the solar system. Some planets and dwarf planets spin backward compared to the Earth, other planets are tipped over, and others orbit outside the *ecliptic* plane by substantial angles, Pluto in particular (see the following figure on *obliquity* and orbital *inclination*). Some planets and moons are still hot enough to be volcanic, and some produce *silicate* lava (for example, the Earth and Jupiter's moon Io), while others have exotic lavas made of molten ices (for example, Neptune's moon Triton).

Today, we look outside our solar system and find planets orbiting other stars, more than 400 to date. Now our search for signs of life goes beyond Mars and Enceladus and Titan and reaches to other star systems. Most of the science presented in this set comes from the startlingly rapid developments of the last 100 years, brought about by technological development.

The rapid advances of planetary and heliospheric science and the astonishing plethora of images sent back by missions motivate the revised editions of the Solar System set. The multivolume set explores the vast and enigmatic Sun at the center of the solar system and moves out through the planets, dwarf planets, and minor bodies of the solar system, examining each and comparing them from the point of view of a planetary scientist. Space missions that produced critical data for the understanding of solar system bodies are introduced in each volume, and their data and images shown and discussed. The revised editions of *The Sun, Mercury, and Venus, The Earth and the Moon,* and *Mars* place emphasis on the areas of unknowns and the results of new space missions. The important fact that the solar system consists of a continuum of sizes and types of bodies is stressed in the revised edition of *Asteroids, Meteorites, and Comets.* This book discusses the roles of these small bodies as recorders of the formation of the solar system, as well as their threat as *impactors* of planets. In the revised edition of *Jupiter and Saturn,* the two largest planets are described and compared. In the revised edition of *Uranus, Neptune, Pluto, and the Outer Solar System,* Pluto is presented in its rightful, though complex, place as the second-largest known of a extensive population of icy bodies that reach far out toward the closest stars, in effect linking the solar system to the Galaxy itself.

This set hopes to change the familiar and archaic litany *Mercury, Venus, Earth, Mars, Jupiter, Saturn, Uranus, Neptune, Pluto* into a thorough understanding of the many sizes and types of bodies that orbit the Sun. Even a cursory study of each planet shows its uniqueness along with the great areas of knowledge that are unknown. These titles seek to make the familiar strange again.

Acknowledgments

foremost, profound thanks to the following organizations for the great science and adventure they provide for humankind and, on a more prosaic note, for allowing the use of their images for these books: the National Aeronautics and Space Administration (NASA) and the National Oceanic and Atmospheric Administration (NOAA), in conjunction with the Jet Propulsion Laboratory (JPL) and Malin Space Science Systems (MSSS). A large number of missions and their teams have provided invaluable data and images, including the Solar and Heliospheric Observer (SOHO), Mars Global Surveyor (MGS), Mars Odyssey, the Mars Exploration Rovers (MERs), Galileo, Stardust, Near-Earth Asteroid Rendezvous (NEAR), and Cassini. Special thanks to Steele Hill, SOHO Media Specialist at NASA, who prepared a number of images from the SOHO mission, to the astronauts who took the photos found at Astronaut Photography of the Earth, and to the providers of the National Space Science Data Center, Great Images in NASA, and the NASA/JPL Planetary Photojournal, all available on the Web (addresses given in the reference section).

Many thanks also to Frank K. Darmstadt, executive editor at Facts On File; to Jodie Rhodes, literary agent; and to E. Marc Parmentier at Brown University for his generous support.

Introduction

Humankind has recognized both Jupiter and Saturn as planets since antiquity. Jupiter, in particular, was revered as a large and powerful presence by ancient peoples. This demonstrates a remarkable level of acuity: Imagine the patience and diligence required to study the night sky and to track the planets without telescopes and or prior knowledge.

Although Jupiter and Saturn were part of the pantheon of gods in antiquity, people began to reform their ideas about the solar system toward the end of the Renaissance in the 17th century. In 1608, Hans Lippershey, a Dutch eyeglassmaker, attached a lens to each end of a hollow tube and created the first telescope. Galileo Galilei (1564–1642), the Italian astronomer born in Pisa, made his first telescope in 1609 from Lippershey's model. Galileo saw that Jupiter had moons orbiting around it. This first observation of moons orbiting a body other than the Earth transformed humankind's perception of the universe and earned Galileo a jail sentence on the charge of being a heretic. However, it could no longer be denied that the Earth was not the center of all heavenly bodies.

Today, new observations of the moons of Jupiter and Saturn are again transforming our vision of the solar system and raising the possibility of life on other planets. The Cassini mission has delivered information about the surface of Saturn's moon Titan, the last surface in the solar system to be seen. Cassini went on to photograph water geysers erupting through the surface of the icy moon Enceladus. These water geysers, unexpected because of the low temperatures on Enceladus, immediately indicate to scientists one critical thing: Enceladus may be a candidate for habitability and is now a major target in scientists' search for life away from the Earth. The

Galileo Galilei discovered the satellites Io, Europa, Ganymede, and Callisto in 1610. They were the first bodies in the solar system that were seen to be obviously orbiting a body other than the Earth. (NASA/ JPL/Galileo)

geysers on Enceladus are so scientifically important that they alone would have motivated a second edition of this book. The fundamental nature of the Cassini mission discoveries, relating to the habitability of the moons of the giant planets and motivating new searches for life in the solar system, requires inclusion in this volume.

Jupiter and Saturn, Revised Edition, is divided into two parts. Part One covers what is known and still unknown about Jupiter, 400 years later in the development of telescopes and astronomy. Chapters 1, 2, and 3 discuss the orbital movements of Jupiter in the solar system and its internal and external composition, movements, and temperatures. Jupiter's moons and rings (just discovered in 1979) are covered in chapter 4. Jupiter's moons have played a critical part in the development of understanding of the solar system: When Galileo discovered four of Jupiter's moons in 1610 they were the first bodies in the solar system that clearly could be demonstrated not to be orbiting Earth. These moons, Io, Europa, Ganymede, and Callisto, are still called the Galilean satellites, or Galilean moons. Their discovery marked the beginning of the end of the wildly incorrect theory that the Earth is the center of the universe and that all heavenly objects orbit this planet. Galileo began making telescopes from Venetian glass and sending them to scientists throughout Europe. He sent a telescope to Johannes Kepler (1571–1630), the prominent German mathematician and astronomer. Kepler did not believe that moons around Jupiter could exist until he saw them with his own eyes. When Kepler saw the moons for himself he promptly coined the word "satellite" from a Latin term meaning "hangers-on to a prominent man."

Saturn, the topic of Part Two, was another major target for the early astronomers and is still at the heart of the most active planetary research. Basic data on the planet along with its orbital dynamics are covered in chapter 7. Saturn's interior structure, heat production, and magnetic field, covered in chapter 8, are all similar to Jupiter's. Saturn's weather and surface, discussed in chapter 9, are overshadowed by dramatic Jupiter, being fainter and harder to see. Saturn's outstanding feature, of course, is its rings. Galileo saw Saturn's immense ring system but did not perceive correctly what he saw, writing in his notebook that Saturn appeared to have handles. Saturn's rings and moons are discussed in chapter 10.

In 1655, Christiaan Huygens (1629–95), a Dutch inventor, mathematician, and astronomer, discovered the first known moon of Saturn. Centuries of additional observation have yielded great results: At the time of writing, 62 moons have been identified, three by the Cassini mission. Within 10 years of Huygens's landmark discovery of the first known moon of Saturn, the banded cloud patterns and the great red spot, that huge cyclonic storm in Jupiter's southern hemisphere, were first seen. Suddenly, scientists were faced with planets that had their own orbiting satellites, immense weather patterns unlike any seen on Earth, and rings around a planet formed by processes they could not yet imagine. The distant giant planets were proving to be entirely unlike the Earth, the Moon, and Mars, and observations were pushing well past the bounds of what scientists could explain.

The excitement of continuous discovery in these planetary systems continues today. Jupiter and Saturn are the two largest planets in the solar system and the first two gas giant planets encountered when moving away from the Sun. They are both immensely complicated systems, miniature solar systems of their own. Jupiter has 63 known moons and its own ring system. Saturn is famous for its rings and also has 62 known moons, which makes it close to overtaking Jupiter as the planet with the largest number of known moons. With some regularity, astronomers are finding new, distant moons

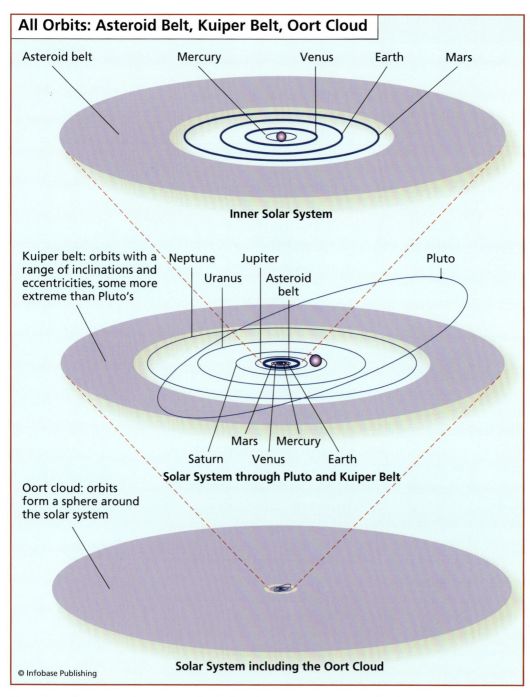

All Orbits: Asteroid Belt, Kuiper Belt, Oort Cloud

Asteroid belt Mercury Venus Earth Mars

Inner Solar System

Kuiper belt: orbits with a range of inclinations and eccentricities, some more extreme than Pluto's

Neptune Jupiter Pluto

Uranus Asteroid belt

Mars Mercury

Saturn Venus Earth

Solar System through Pluto and Kuiper Belt

Oort cloud: orbits form a sphere around the solar system

© Infobase Publishing

Solar System including the Oort Cloud

The orbits of Jupiter and Saturn are highlighted here. All the orbits are far closer to circular than shown in this oblique view, which was chosen to show the inclination of Pluto's orbit to the ecliptic.

orbiting each of these planets, so their number of known satellites will certainly continue to grow.

Jupiter is the largest planet in the solar system, but it contains only 1/10 of 1 percent of the mass of the Sun. If it had been only 80 times larger, it would have had the mass required to begin nuclear fusion in its interior, and this solar system would have had two stars. By comparison, the Earth would have had to be 27,000 times more massive to become a star. Saturn has the youngest and freshest ring system in the solar system, presenting a beautiful laboratory for the study and understanding of rings. Each of these gas giants has a moon that has come almost to eclipse the scientific interest of the body itself: Saturn's largest moon, Titan, has an atmosphere like that of early Earth, and Jupiter's moon Europa has a water ocean and thus is the most likely place in the solar system to find life now. (Mars may have had life in the past but is less likely to have life now.)

Throughout this book, the two planets will be described as they are known now, but, more important, the directions for current research and the outstanding questions will be presented. Saturn's rings present a model for the early solar nebula, and Titan's atmosphere gives an example for Earth before the development of life. Io, one of Jupiter's large moons, is the most volcanically active body in the solar system (and the only body beside Earth on which man has witnessed volcanic activity) and provides a model for the massive flood *basalts* that have disrupted Earth's climate in the past. A remote possibility exists that the moon Europa may harbor life in the present. Through these, and many more unusual aspects of these giant planetary systems, Earth's past, its present, and possibly its future can be studied.

PART ONE

JUPITER

Jupiter: Fast Facts about a Planet in Orbit

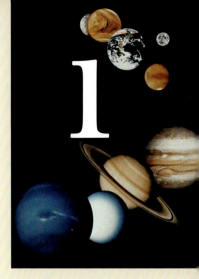

Jupiter is one of the brightest objects in the sky, despite its great distance from Earth. The brightness of a celestial object when seen from a given distance is called its *apparent magnitude*. This scale has no dimensions, but allows comparison between objects. The lower the magnitude number, the brighter the object. The full Moon has magnitude -12.7 and the Sun has -26.7. The faintest stars visible under dark skies are around +6. Jupiter has an apparent magnitude of about -2. Mars is about as bright as Jupiter, and all the other planets are dimmer than the brightest star visible from Earth, which is Sirius, at apparent magnitude -1.46 (Saturn's apparent magnitude is only +0.6).

Jupiter's brightness has made it an enticing target for astronomers for centuries. Galileo discovered the first of its moons, and Giovanni (Jean-Dominique) Cassini, another highly productive Italian astronomer, discovered the planet's bands and the Great Red Spot around 1655. (For more on the amazing Cassini, see the sidebar "Giovanni Cassini" on page 152.) Other sources report that Robert Hooke, the great English experimentalist, was the first to see the Great Red Spot, in 1664. In 1664 Cassini made another important observation of Jupiter: It is slightly flattened at its poles and bulges at

Jupiter's complex and changing cloud patterns have made it a compelling target for astronomers since the invention of the telescope in 1608. (NASA/JPL)

its equator. Cassini was exactly right in this observation, and though the effect is extreme on Jupiter, in fact all planets are slightly flattened. The flattening cannot be seen easily in the image of Jupiter.

Each planet and some other bodies in the solar system (the Sun and certain asteroids) have been given their own symbol as a shorthand in scientific writing. The symbol for Jupiter is shown on page 6.

Jupiter's spin causes its polar flattening and equatorial bulge. A planet's rotation prevents it from being a perfect sphere. Spinning around an axis creates forces that cause the planet to swell at the equator and flatten slightly at the poles. Planets are thus shapes called oblate spheroids, meaning that

FUNDAMENTAL INFORMATION ABOUT JUPITER

Along with being the brightest of all planets, Jupiter is by far the most massive. Despite its low density, just higher than that of liquid water on Earth, it is almost four times as massive as its neighbor Saturn, and more than 300 times as massive as the Earth (Saturn's density, strangely, is just more than half of Jupiter's, and is in fact less than the density of water). If Jupiter had been just 15 times more massive nuclear fusion would have begun in its *core,* and this solar system would have had two stars rather than just one.

Jupiter also excels in its number of known moons (63, slightly more than Saturn, the next contender) and the largest magnetic field of any planet. Jupiter's huge magnetic field engulfs many of its moons, protecting them from the damages of the solar wind. The table below lists a number of Jupiter's physical characteristics.

FUNDAMENTAL FACTS ABOUT JUPITER	
equatorial radius at the height where atmospheric pressure is one bar	44,424 miles (71,492 km), 11.21 times Earth's radius
polar radius at the height where atmospheric pressure is one bar	41,490 miles (66,770 km)
ellipticity	0.066, meaning the planet's equator is about 6 percent longer than its polar radius
volume	3.42×10^{14} cubic miles (1.43×10^{15} km³), or 1,316 times Earth's volume
mass	4.2×10^{27} pounds (1.9×10^{27} kg), or 317.8 times Earth's mass
average density	83 pounds per cubic foot (1,330 kg/m³)
acceleration of gravity at the equator and at the altitude where atmospheric pressure is one bar	75.57 feet per second squared (23.12 m/sec²), or 2.36 times Earth's
magnetic field strength at the surface	4.2×10^{-4} tesla, or about 10 times Earth's
rings	3
moons	63 presently known

they have different equatorial radii and polar radii, as shown in the image here. If the planet's equatorial radius is called r_e, and its polar radius is called r_p, then its flattening (more commonly called ellipticity, e, shown in the figure on page 7) is defined as

$$e = \frac{r_e - r_p}{r_e}$$

The larger radius, the equatorial, is also called the *semimajor axis*, and the polar radius is called the *semiminor axis*. Jupiter's semimajor axis is 44,424 miles (71,492 km), and its semiminor axis is 41,490 miles (66,770 km), so its ellipticity (see figure on page 7) is

$$e = \frac{71,492 - 66,770}{71,492} = 0.066.$$

Because every planet's equatorial radius is longer than its polar radius, the surface of the planet at its equator is farther from the planet's center than the surface of the planet at

Symbol for Jupiter

© Infobase Publishing

Many solar system objects have simple symbols—this is the symbol for Jupiter.

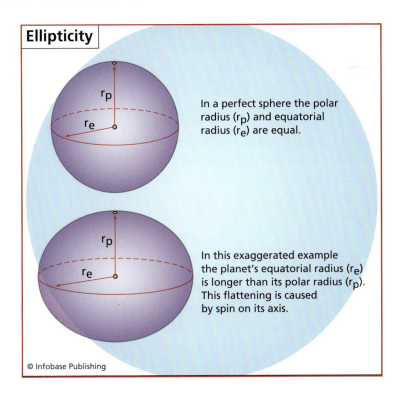

Ellipticity

In a perfect sphere the polar radius (r_p) and equatorial radius (r_e) are equal.

In this exaggerated example the planet's equatorial radius (r_e) is longer than its polar radius (r_p). This flattening is caused by spin on its axis.

© Infobase Publishing

Ellipticity is the measure of how much a planet's shape deviates from a sphere.

its poles. Being at a different distance from the center of the planet means there is a different amount of mass between the surface and the center of the planet. Mass pulls on nearby objects with gravity. (For more information on gravity, see the sidebar "What Makes Gravity?" on page 9.) At the equator, where the radius of the planet is larger and the amount of mass beneath is therefore relatively larger, the pull of gravity is actually stronger than it is at the poles. Gravity is therefore not a perfect constant on any planet: Variations in radius, topography, and the density of the material underneath make the gravity vary slightly over the surface. This is why planetary gravitational accelerations are generally given as an average value on the planet's equator, and in a gas giant planet, gravity is given as an average at the height where pressure equals one atmosphere.

Just as planets are not truly spheres, the orbits of solar system objects are not circular. Johannes Kepler, the prominent 17th-century German mathematician and astronomer, first realized that the orbits of planets are ellipses after analyzing a series of precise observations of the location of Mars that had been taken by his colleague, the distinguished Danish astronomer Tycho Brahe. Kepler drew rays from the Sun's center to the orbit of Mars, and noted the date and time that Mars arrived on each of these rays. He noted that Mars swept out equal areas between itself and the Sun in equal times, and that Mars moved much faster when it was near the Sun than when it was farther from the Sun. Together, these observations convinced Kepler that the orbit was shaped as an ellipse, and not as a circle, as had been previously assumed. Kepler defined three laws of orbital motion (listed in the table on page 10), which he published in 1609 and 1619 in his books *New Astronomy* and *The Harmony of the World*. These three laws are still used as the basis for understanding orbits.

As Kepler observed, all orbits are ellipses, not circles. An ellipse can be thought of simply as a squashed circle,

The ellipticities of the planets differ largely as a function of their composition's ability to flow in response to rotational forces.

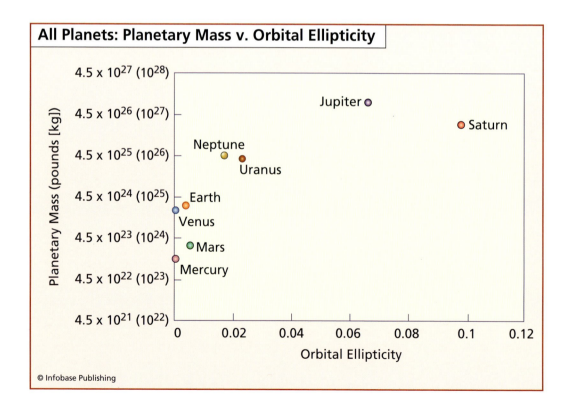

All Planets: Planetary Mass v. Orbital Ellipticity

Planetary Mass (pounds [kg])

4.5×10^{27} (10^{28})
4.5×10^{26} (10^{27})
4.5×10^{25} (10^{26})
4.5×10^{24} (10^{25})
4.5×10^{23} (10^{24})
4.5×10^{22} (10^{23})
4.5×10^{21} (10^{22})

Jupiter
Saturn
Neptune
Uranus
Earth
Venus
Mars
Mercury

0   0.02   0.04   0.06   0.08   0.1   0.12

Orbital Ellipticity

© Infobase Publishing

WHAT MAKES GRAVITY?

Gravity is among the least understood forces in nature. It is a fundamental attraction between all matter, but it is also a very weak force: The gravitational attraction of objects smaller than planets and moons is so weak that electrical or magnetic forces can easily oppose it. At the moment about the best that can be done with gravity is to describe its action: How much mass creates how much gravity? The question of what makes gravity itself is unanswered. This is part of the aim of a branch of mathematics and physics called string theory: to explain the relationships among the natural forces and to explain what they are in a fundamental way.

Sir Isaac Newton, the English physicist and mathematician who founded many of today's theories back in the mid-17th century, was the first to develop and record universal rules of gravitation. There is a legend that he was hit on the head by a falling apple while sitting under a tree thinking, and the fall of the apple under the force of Earth's gravity inspired him to think of matter attracting matter.

The most fundamental description of gravity is written in this way:

$$F = \frac{Gm_1 m_2}{r^2},$$

where F is the force of gravity, G is the universal gravitational constant (equal to 6.67 $\times$ 10^{-11} Nm^2/kg^2), m_1 and m_2 are the masses of the two objects that are attracting each other with gravity, and r is the distance between the two objects. (N is the abbreviation for newtons, a metric unit of force.)

Immediately, it is apparent that the larger the masses, the larger the force of gravity. In addition, the closer together they are, the stronger the force of gravity, and because r is squared in the denominator, gravity diminishes very quickly as the distance between the objects increases. By substituting numbers for the mass of the Earth (5.9742 $\times$ 10^{24} kg), the mass of the Sun (1.989 $\times$ 10^{30} kg), and the distance between them, the force of gravity between the Earth and Sun is shown to be 8 $\times$ 10^{21} pounds per foot (3.56 $\times$ 10^{22} N). This is the force that keeps the Earth in orbit around the Sun. By comparison, the force of gravity between a piano player and her piano when she sits playing is about 6 $\times$ 10^{-7} pounds per feet (2.67 $\times$ 10^{-6} N). The force of a pencil pressing down in the palm of a hand under the influence of Earth's gravity is about 20,000 times stronger than the gravitational attraction between the player and the piano! So, although the player and the piano are attracted to each other by gravity, their masses are so small that the force is completely unimportant.

KEPLER'S LAWS

Kepler's first law:	A planet orbits the Sun following the path of an ellipse with the Sun at one focus.
Kepler's second law:	A line joining a planet to the Sun sweeps out equal areas in equal times (see figure below).
Kepler's third law:	The closer a planet is to the Sun, the greater its speed. This is stated as: The square of the period of a planet T is proportional to the cube of its semimajor axis R, or $T \propto R^{\frac{3}{2}}$, as long as T is in years and R in AU.

Kepler's Second Law

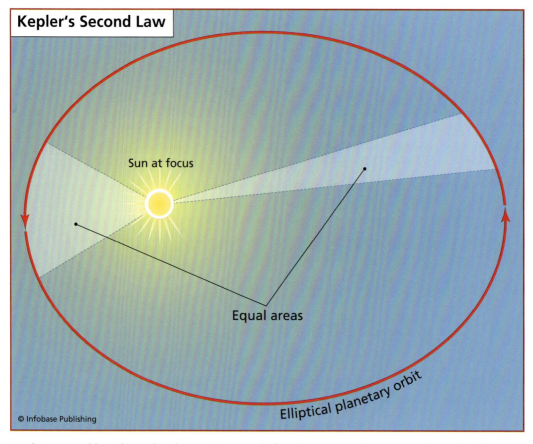

Sun at focus

Equal areas

Elliptical planetary orbit

© Infobase Publishing

Kepler's second law shows that the varying speed of a planet in its orbit requires that a line between the planet and the Sun sweep out equal areas in equal times.

resembling an oval. The proper definition of an ellipse is the set of all points that have the same sum of distances to two given fixed points, called foci. To demonstrate this definition, take two pins, push them into a piece of stiff cardboard, and loop a piece of string around the pins (see figure below). The two pins are the foci of the ellipse. Pull the string away from the pins with a pencil, and draw the ellipse, keeping the string taut around the pins and the pencil all the way around. Adding the distance along the two string segments from the pencil to each of the pins will give the same answer each time: The ellipse is the set of all points that have the same sum of distances from the two foci.

The mathematical equation for an ellipse is

$$\frac{x^2}{a^2} + \frac{y^2}{b^2} = 1,$$

where x and y are the coordinates of all the points on the ellipse, and a and b are the semimajor and semiminor axes, respectively. The semimajor axis and semiminor axis would both be the radius if the shape was a circle, but two radii are

Making an ellipse with string and two pins— adding the distance along the two string segments from the pencil to each of the pins will give the same sum at every point around the ellipse. This method creates an ellipse with the pins at its foci.

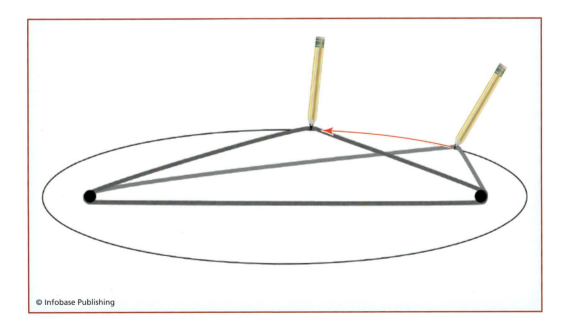

needed for an ellipse. If a and b are equal, then the equation for the ellipse becomes the equation for a circle:

$$x^2 + y^2 = n,$$

where n is any constant.

When drawing an ellipse with string and pins, it is obvious where the foci are (they are the pins). In the abstract, the foci can be calculated according to the following equations:

Coordinates of the first focus

$$= \left(+\sqrt{a^2 - b^2}, 0 \right)$$

Coordinates of the second focus

$$= \left(-\sqrt{a^2 - b^2}, 0 \right)$$

In the case of an orbit, the object being orbited (for example, the Sun) is located at one of the foci (see figure below).

An important characteristic of an ellipse, perhaps the most important for orbital physics, is its eccentricity: a measure of how different the semimajor and semiminor axes of the ellipse are. Eccentricity is dimensionless and ranges from 0

The semimajor and semiminor axes of an ellipse (or an orbit) are the elements used to calculate its eccentricity, and the body being orbited always lies at one of the foci.

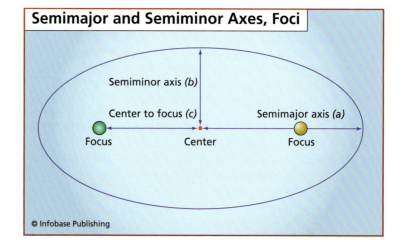

Semimajor and Semiminor Axes, Foci

Semiminor axis (b)

Center to focus (c)

Semimajor axis (a)

Focus

Center

Focus

© Infobase Publishing

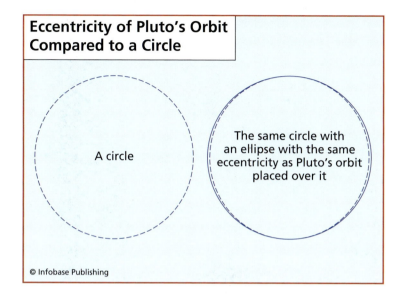

Eccentricity of Pluto's Orbit Compared to a Circle

A circle

The same circle with an ellipse with the same eccentricity as Pluto's orbit placed over it

© Infobase Publishing

Though the orbits of planets are measurably eccentric, they deviate from circularity by very little. This figure shows the eccentricity of Pluto's orbit in comparison with a circle.

to 1, where an eccentricity of zero means that the figure is a circle, and an eccentricity of 1 means that the ellipse has gone to its other extreme, a parabola (the reason an extreme ellipse becomes a parabola results from its definition as a conic section). One equation for eccentricity is

$$e = \sqrt{1 - \frac{b^2}{a^2}},$$

where a and b are the semimajor and semiminor axes, respectively. Another equation for eccentricity is

$$e = \frac{c}{a},$$

where c is the distance between the center of the ellipse and one focus. The eccentricities of the orbits of the planets vary widely, though most are very close to circles, as shown in the figure above. Pluto has the most eccentric orbit (0.244), and Mercury's orbit is also very eccentric, but the rest have eccentricities below 0.09.

The table on page 14 lists a number of characteristics of Jupiter's orbit. The planet rotates far faster than the Earth, and lies more than twice as far from the Sun as its inner neighbor Mars. Its orbit is highly regular: Its path is almost circular and tipped very little from the ecliptic plane. Jupiter's rotation axis is similarly almost perfectly perpendicular to its orbital plane, in contrast to planets like the Earth, Mars, and Saturn, which have significant axial tilts (obliquities) that cause these planets to have seasons.

While the characteristics of an ellipse drawn on a sheet of paper can be measured, orbits in space are more difficult to characterize. The ellipse itself has to be described, then the ellipse's position in space, and then the motion of the body as it travels around the ellipse. Six parameters are needed to specify the motion of a body in its orbit and the position of the

JUPITER'S ORBIT	
rotation on its axis ("day")	9.9 Earth hours, but varies from equator to poles, promoting atmospheric mixing
rotation speed at equator rotation direction	28,122 MPH (45,259 km/hour) prograde (counterclockwise when viewed from above the North Pole)
sidereal period ("year")	11.86 Earth years
orbital velocity (average)	8.123 miles per second (13.07 m/sec)
sunlight travel time (average)	43 minutes and 16 seconds to reach Jupiter
average distance from the Sun	483,696,023 miles (778,412,010 km), or 5.2 AU
perihelion	460,276,100 miles (740,742,600 km), or 4.952 AU from the Sun
aphelion	507,089,500 miles (816,081,400 km), or 5.455 AU from the Sun
orbital eccentricity	0.04839
orbital inclination to the ecliptic	1.304 degrees
obliquity (inclination of equator to orbit)	3.12 degrees

orbit. These are called the orbital elements (see the figure on page 18). The first three elements are used to determine where a body is in its orbit.

a **semimajor axis** The semimajor axis is half the width of the widest part of the orbit ellipse. For solar system bodies, the value of the semimajor axis is typically expressed in units of AU. Jupiter's semimajor axis is 5.2044 AU.

e **eccentricity** Eccentricity measures the amount by which an ellipse differs from a circle, as described above. An orbit with $e = 0$ is circular, and an orbit with $e = 1$ stretches into infinity and becomes a parabola. In between, the orbits are ellipses. The orbits of all large planets are almost circles: The Earth, for instance, has an eccentricity of 0.0068, and Jupiter's eccentricity is 0.04839.

M **mean anomaly** Mean anomaly is an angle that moves in time from 0 to 360 degrees during one revolution, as if the planet were at the end of a hand of a clock and the Sun were at its center. This angle determines where in its orbit a planet is at a given time, and is defined to be 0 degrees at *perigee* (when the planet is closest to the Sun) and 180 degrees at *apogee* (when the planet is farthest from the Sun). The equation for mean anomaly M is given as

$$M = M_0 + 360\left(\frac{t}{T}\right),$$

where M_0 is the value of M at time zero, T is the *orbital period,* and t is the time in question.

The next three Keplerian elements determine where the orbit is in space.

i **inclination** For the case of a body orbiting the Sun, the inclination is the angle between the plane of the orbit of the body and the plane of the ecliptic (the plane in which the Earth's orbit lies). For the case of a body orbiting the Earth, the inclination is the angle between the plane of the body's orbit and the plane of the Earth's equator, such that an inclination of zero indicates that the body orbits directly over the equator, and an inclination of 90 indicates that the body orbits over the Poles. If there is an orbital inclination greater than zero, then there is a line of intersection between the ecliptic plane and the orbital plane. This line is called the line of nodes. Jupiter's orbital inclination is 1.304 degrees (see table on page 17).

Ω **longitude of the ascending node** After inclination is specified, there are still an infinite number of orbital planes possible: The line of nodes could cut through the Sun at any longitude around the Sun. Notice that the line of nodes emerges from the Sun in two places. One is called the ascending node (where the orbiting planet crosses the Sun's equator going from south to north). The other is called the descending node (where the orbiting planet crosses the Sun's equator going from north to south). Only one node needs to be specified, and by convention the ascending node is used. A second point in a planet's orbit is the vernal *equinox,* the spring day in which day and night have the same length (the word *equinox* means equal night), occurring where the plane of the planet's equator intersects its orbital plane. The angle between the vernal equinox γ and the ascending node N is called the longitude of the ascending node. Jupiter's longitude of the ascending node is 100.45 degrees.

ω **argument of the perigee** The argument of the perigee is the angle (in the body's orbit plane) between the ascending node N and perihelion P, measured in the direction of the body's orbit. Jupiter's argument of the perigee is 14.7539 degrees.

The complexity of the six measurements shown in the figure on page 18 demonstrates the extreme attention to detail

OBLIQUITY, ORBITAL INCLINATION, AND ROTATIONAL DIRECTION FOR ALL THE PLANETS

Planet	Obliquity (inclination of the planet's equator to its orbit; tilt); remarkable values are in italic	Orbital inclination to the ecliptic (angle between the planet's orbital plane and the Earth's orbital plane); remarkable values are in italic	Rotational direction
Mercury	0° (though some scientists believe the planet is flipped over, so this value may be 180°)	7.01°	prograde
Venus	*177.3°*	3.39°	retrograde
Earth	23.45°	0° (by definition)	prograde
Mars	25.2°	1.85°	prograde
Jupiter	3.12°	1.30°	prograde
Saturn	26.73°	2.48°	prograde
Uranus	*97.6°*	0.77°	retrograde
Neptune	29.56°	1.77°	prograde
Pluto (now classified as a dwarf planet)	*122.5°*	*17.16°*	retrograde

that is necessary when moving from simple theory ("every orbit is an ellipse") to measuring the movements of actual orbiting planets. Because of the gradual changes in orbits over time caused by gravitational interactions of many bodies and by changes within each planet, natural orbits are complex, evolving motions. To plan with such accuracy space missions such as the recent Mars Exploration Rovers, each of which landed perfectly in their targets, just kilometers long on the surface of another planet, the mission planners must be masters of orbital parameters.

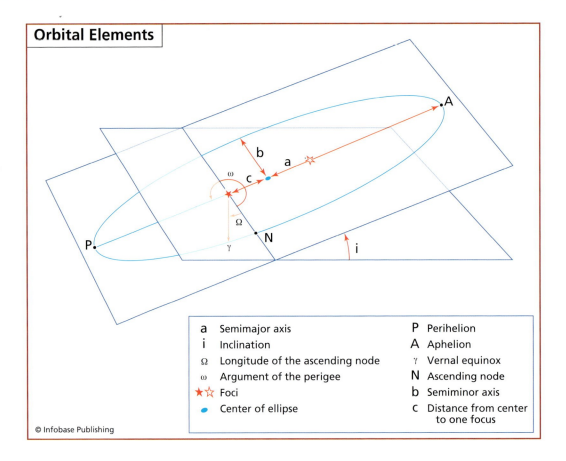

Orbital Elements

a	Semimajor axis	P	Perihelion
i	Inclination	A	Aphelion
Ω	Longitude of the ascending node	γ	Vernal equinox
ω	Argument of the perigee	N	Ascending node
★☆	Foci	b	Semiminor axis
●	Center of ellipse	C	Distance from center to one focus

© Infobase Publishing

A series of parameters called orbital elements are used to describe exactly the orbit of a body.

Though Jupiter and Saturn have immense gravity fields that tend to pull small bodies like asteroids and comets out of their own orbits and cause them to crash on the giant planets, there are safe, stable orbits near the giant planets for small bodies. Joseph-Louis Lagrange, a famous French mathematician who lived in the late 18th and early 19th centuries, calculated that there are five positions in an orbiting system of two large bodies in which a third small body, or collection of small bodies, can exist without being thrown out of orbit by gravitational forces. More precisely, the Lagrange points mark positions where the gravitational pull of the two large bodies precisely equals the centripetal force required to rotate with them. In the solar system the two large bodies are the Sun and a planet, and the smaller body or group of bodies, asteroids. Of the five Lagrange points, three are unstable over long periods, and two are permanently stable. The unstable Lagrange points, L1, L2 and L3, lie along the line connecting the two large masses. The stable Lagrange points, L4 and L5, lie in the orbit of the planet, 60 degrees ahead and 60 degrees behind the planet itself (see figure below).

The asteroids that orbit in the Lagrange points of Jupiter are called Trojan asteroids. Asteroid 624 Hektor orbits 60 degrees ahead of Jupiter, and was the first Trojan asteroid to be discovered, in 1907. Asteroid 624 Hektor is 186 × 93 miles (300 × 150 km), the largest of all the Trojans. There are about 1,200 Trojan asteroids now known. These asteroids are named after the Greek besiegers of Troy in the Trojan war, and for the Trojan opponents. As these asteroids are discovered, they are named for the men in the battles: Those at the Lagrange point L4 are all meant to be named after Greeks, and those at L5 after Trojans. Unfortunately a few errors have been made, placing people on the wrong side of the battle, and some astronomers affectionately refer to these misnamed asteroids as spies. Asteroid 617 Patroclus, for example, named

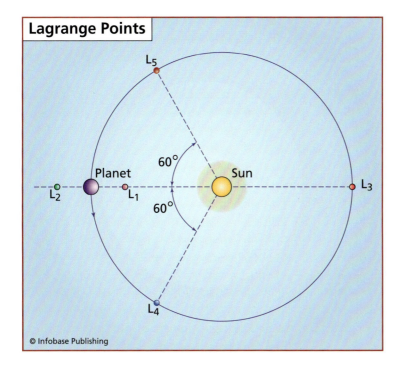

Lagrange Points

© Infobase Publishing

There are five positions, called Lagrange points, relative to Jupiter in which small bodies can orbit stably. Jupiter's Lagrange points L4 and L5 hold collections of asteroids called the Trojans.

after a prominent Greek hero, is in the L5 Trojan Lagrange point.

Jupiter orbits the Sun in a highly regular way, with an almost-circular orbit little inclined to the ecliptic, and with its rotation axis almost perpendicular to its orbital plane. Jupiter itself forms a kind of miniature solar system: It has a collection of more than 60 moons, from large, volcanically active moons (Io in particular) to a far-flung population of captured asteroids, and even shares its orbit with two populations of Trojan asteroids. This solar system's largest planet and most complex planetary system is almost a miniature solar system in itself.

The Interior of Jupiter

Having moved past Mars in another volume of The Solar System set, we now encounter planets with entirely different appearances and behaviors. Jupiter is the first of the gas giant planets, as different from its inner neighbor Mars as it is possible to be in this solar system. Where Mars is a small rocky planet with an iron-rich metallic core, like Earth, Jupiter is an immense sphere of gaseous, liquid, and metallic hydrogen and helium. Mars has lost its magnetic field, but Jupiter has the most powerful magnetic field in the solar system aside from the Sun's. The interior states and processes of Jupiter and the other gas giants are entirely different from those of the *terrestrial planets*.

MATERIALS, TEMPERATURES, AND PRESSURES

Gaseous planets consist of gas and fluid to a very great depth in their interiors. At the depth in Jupiter where the pressure is about one bar (~1 atm), there are clouds of frozen ammonia and hydrogen sulfide, which appear brown from the Earth. Below the frozen clouds there is probably a layer of thick fluid clouds (like the clouds on Earth, made of tiny fluid drops instead of ice crystals). Though a commonly chosen pressure for a defined surface is one bar, the average atmospheric pressure at sea level on Earth, on Jupiter the surface is considered to be below the cloudbanks, at a pressure of 10 bars.

Jupiter's atmosphere can be penetrated by visible and infrared radiation into the *troposphere* (the deepest part of the atmosphere, where temperature decreases with altitude), down to pressures of a few bars. The upper *stratosphere* (the layer above the troposphere, where temperature increases with altitude) can also be probed at specific infrared wavelengths that allow the identification of specific ions. (For more information, see the sidebar "Elements and Isotopes" on page 23.) Direct information about deeper layers of Jupiter has also been made available by the *Galileo* probe, which penetrated Jupiter to a pressure of about 20 bars. This probe entered a clear zone, thought to be dry and hot compared to the cloudy belts of the planet, and so its data may not be representative of the entire atmosphere.

In these ways compositions and temperatures of the atmosphere to depths corresponding to a few bars in pressure have been directly measured. At greater depths, the internal structures of Jupiter are only known from indirect measurements and theoretical modeling.

Jupiter is thought to consist primarily of primordial gases from the protoplanetary disk, mostly hydrogen and helium. Its central core is probably rock and ice with a mass of about 10 to 30 Earth masses. The entire cloud zone that is seen in images of Jupiter is extremely thin, only about 31 miles (50 km), and below it lies a clear atmosphere mainly of gaseous helium and hydrogen. As pressures increase inside the planet with depth, the helium and hydrogen gradually become liquid (see figure on page 26).

Pressures inside the planet finally become high enough to press the hydrogen into a metallic liquid in which electrons move freely among the hydrogen nuclei. (This assertion is based on modeling and experimentation, not on direct observations.) This abrupt transition to hydrogen and helium metal occurs at about 0.78 of the radius of the planet. The hydrogen metal phase is not hard and stiff, as metal is on Earth's surface. This substance is hot and moves in currents like a thick

ELEMENTS AND ISOTOPES

All the materials in the solar system are made of *atoms* or of parts of atoms. A family of atoms that has the same number of positively charged particles in their nuclei (the center of the atom) is called an *element:* Oxygen and iron are elements, as are aluminum, helium, carbon, silicon, platinum, gold, hydrogen, and well over 200 others. Every single atom of oxygen has eight positively charged particles, called protons, in its nucleus. The number of protons in an atom's nucleus is called its *atomic number:* All oxygen atoms have an atomic number of 8, and that is what makes them all oxygen atoms.

Naturally occurring nonradioactive oxygen, however, can have either eight, nine, or 10 uncharged particles, called neutrons, in its nucleus, as well. Different weights of the same element caused by addition of neutrons are called *isotopes.* The sum of the protons and neutrons in an atom's nucleus is called its *mass number.* Oxygen can have mass numbers of 16 (eight positively charged particles and eight uncharged particles), 17 (eight protons and nine neutrons), or 18 (eight protons and 10 neutrons). These isotopes are written as ^{16}O, ^{17}O, and ^{18}O. The first, ^{16}O, is by far the most common of the three isotopes of oxygen.

Atoms, regardless of their isotope, combine together to make molecules and compounds. For example, carbon (C) and hydrogen (H) molecules combine to make methane, a common gas constituent of the outer planets. Methane consists of one carbon atom and four hydrogen atoms and is shown symbolically as CH_4. Whenever a subscript is placed by the symbol of an element, it indicates how many of those atoms go into the makeup of that molecule or compound.

Quantities of elements in the various planets and moons, and ratios of isotopes, are important ways to determine whether the planets and moons formed from the same material or different materials. Oxygen again is a good example. If quantities of each of the oxygen isotopes are measured in every rock on Earth and a graph is made of the ratios of $^{17}O/^{16}O$ versus $^{18}O/^{16}O$, the points on the graph will form a line with a certain slope (the slope is 1/2, in fact). The fact that the data forms a line means that the material that formed the Earth was homogeneous; beyond rocks, the oxygen isotopes in every living thing and in the atmosphere also lie on this slope. The materials on the Moon also show this same slope. By measuring oxygen isotopes in many different kinds of solar system materials, it has now been shown that the slope of the plot $^{17}O/^{16}O$

(continues)

(continued)

versus $^{18}O/^{16}O$ is one-half for every object, but each object's line is offset from the others by some amount. Each solar system object lies along a different parallel line.

At first it was thought that the distribution of oxygen isotopes in the solar system was determined by their mass: The more massive isotopes stayed closer to the huge gravitational force of the Sun, and the lighter isotopes strayed farther out into the solar system. Studies of very primitive meteorites called chondrites, thought to be the most primitive, early material in the solar system, showed to the contrary that they have heterogeneous oxygen isotope ratios, and therefore oxygen isotopes were not evenly spread in the early solar system. Scientists then recognized that temperature also affects oxygen isotopic ratios: At different temperatures, different ratios of oxygen isotopes condense. As material in the early solar system cooled, it is thought that first aluminum oxide condensed, at a temperature of about 2,440°F (1,340°C), and then calcium-titanium oxide ($CaTiO_3$), at a temperature of about 2,300°F (1,260°C), and then a calcium-aluminum-silicon-oxide ($Ca_2Al_2SiO_7$), at a temperature of about 2,200°F (1,210°C), and so on through other compounds down to iron-nickel alloy at 1,800°F (990°C) and water, at -165°F (-110°C) (this low temperature for the condensation of water is caused by the very low pressure of space). Since oxygen isotopic ratios vary with temperature, each of these oxides would have a slightly different isotopic ratio, even if they came from the same place in the solar system.

The key process that determines the oxygen isotopes available at different points in the early solar system nebula seems to be that simple compounds created with ^{18}O

liquid. (For more, see the sidebar "Rheology, or How Solids Can Flow" on page 106.) The last compositional layer in Jupiter is thought to be an ice-silicate core, taking up only the inner 20 percent of the planet's radius.

While pressures are low enough to allow the hydrogen and helium to be gases or liquids, they freely intermix. When pressures increase, (see the sidebar "What Is Pressure?" on page 27) hydrogen is pressed into a metal; however, helium cannot fit into the hydrogen metal structure and separates into distinct droplets. The helium droplets are heavier than the hydrogen metal, and so they sink further into Jupiter's interior, as a sort of helium rain. This sinking releases energy that reheats the interior. As less dense helium gas transforms

are relatively stable at high temperatures, while those made with the other two isotopes break down more easily and at lower temperatures. Some scientists therefore think that ^{17}O and ^{18}O were concentrated in the middle of the nebular cloud, and ^{16}O was more common at the edge. Despite these details, though, the basic fact remains true: Each solar system body has its own slope on the graph of oxygen isotope ratios.

Most atoms are stable. A carbon-12 atom, for example, remains a carbon-12 atom forever, and an oxygen-16 atom remains an oxygen-16 atom forever, but certain atoms eventually disintegrate into a totally new atom. These atoms are said to be "unstable" or "radioactive." An unstable atom has excess internal energy, with the result that the nucleus can undergo a spontaneous change toward a more stable form. This is called "radioactive decay." Unstable isotopes (radioactive isotopes) are called "radioisotopes." Some elements, such as uranium, have no stable isotopes. The rate at which unstable elements decay is measured as a "half-life," the time it takes for half of the unstable atoms to have decayed. After one half-life, half the unstable atoms remain; after two half-lives, one-quarter remain, and so forth. Half-lives vary from parts of a second to millions of years, depending on the atom being considered. Whenever an isotope decays, it gives off energy, which can heat and also damage the material around it. Decay of radioisotopes is a major source of the internal heat of the Earth today: The heat generated by accreting the Earth out of smaller bodies and the heat generated by the giant impactor that formed the Moon have long since conducted away into space.

to liquid and liquid to dense metal beneath the hydrogen layer, the planet becomes denser without changing its mass. This transformation may contribute to the tiny, ongoing contraction of the planet. The helium rain is a hypothesis arrived at from theory, since it cannot be observed directly. The descent of helium into the deep planet, however, should produce a lack of helium at the surface in comparison to solar values. This prediction is in fact observed at the surface: *Galileo* observed a reduction in helium in the surface of Jupiter, compared to expected solar system values. There is also a deficit in neon, perhaps because neon can dissolve into the helium droplet and be carried into the deep interior as well.

Internal Structure of Jupiter

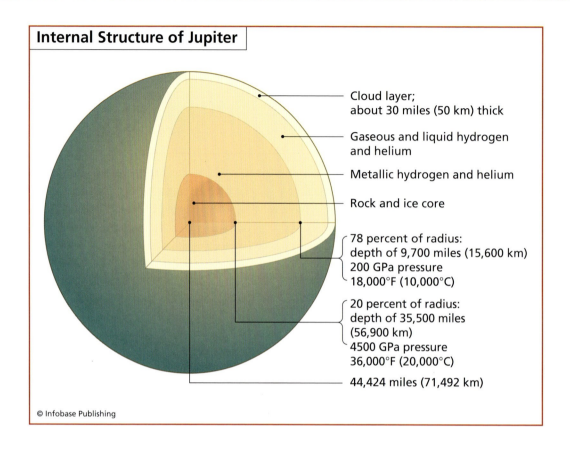

Cloud layer;
about 30 miles (50 km) thick

Gaseous and liquid hydrogen
and helium

Metallic hydrogen and helium

Rock and ice core

78 percent of radius:
depth of 9,700 miles (15,600 km)
200 GPa pressure
18,000°F (10,000°C)

20 percent of radius:
depth of 35,500 miles
(56,900 km)
4500 GPa pressure
36,000°F (20,000°C)

44,424 miles (71,492 km)

© Infobase Publishing

The internal structure of Jupiter beneath the cloud layer is described using theory, since there are no direct measurements.

At Jupiter's surface the radius at which pressure is about 10 bars (10^{-3} GPa), the temperature is -148°F (-100°C). Between the surface and just 22 percent of the way down into the planet, where the molecular hydrogen and helium become metallic, the temperature has risen to 18,000°F (10,000°C), and the pressure to 1,974,000 atm (200 GPa), 10^5 times the 10 bar surface pressure. At the boundary of Jupiter's core, 80 percent of the way to the planet's center, temperatures are thought to reach 36,000°F (20,000°C).

When smaller bodies, often called *planetesimals,* are colliding and sticking together, creating a single larger body (perhaps a planet), they are said to be accreting. Eventually the larger body may even have enough gravity itself to begin altering the paths of passing planetesimals and attracting them to it. In any case, the process of *accretion* adds tremendous heat

WHAT IS PRESSURE?

The simple definition of pressure *(p)* is that it is force *(F)* per area *(a)*:

$$p = \frac{F}{a}.$$

Atmospheric pressure is the most familiar kind of pressure and will be discussed below. Pressure, though, is something felt and witnessed all the time, whenever there is a force being exerted on something. For example, the pressure that a woman's high heel exerts on the foot of a person she stands on is a force (her body being pulled down by Earth's gravity) over an area (the area of the bottom of her heel). The pressure exerted by her heel can be estimated by calculating the force she is exerting with her body in Earth's gravity (which is her weight, here guessed at 130 pounds, or 59 kg, times Earth's gravitational acceleration, 32 ft/sec², or 9.8 m/sec²) and dividing by the area of the bottom of the high heel (here estimated as one square centimeter):

$$p = \frac{(59kg)(9.8m / \sec^2)}{(0.01^2 m^2)} = 5,782,000 kg / m\sec^2.$$

The resulting unit, kg/ms², is the same as N/m and is also known as the pascal (Pa), the standard unit of pressure (see appendix 1, "Units and Measurements," to understand more). Although here pressure is calculated in terms of pascals, many scientists refer to pressure in terms of a unit called the atmosphere. This is a sensible unit because one atmosphere is approximately the pressure felt from Earth's atmosphere at sea level, though of course weather patterns cause continuous fluctuation. (This fluctuation is why weather forecasters say "the barometer is falling" or "the barometer is rising:" The measurement of air pressure in that particular place is changing in response to moving masses of air, and these changes help indicate the weather that is to come.) There are about 100,000 pascals in an atmosphere, so the pressure of the woman's high heel is about the same as 57.8 times atmospheric pressure.

What is atmospheric pressure, and what causes it? Atmospheric pressure is the force the atmosphere exerts by being pulled down toward the planet by the planet's

(continues)

(continued)

gravity, per unit area. As creatures of the Earth's surface, human beings do not notice the pressure of the atmosphere until it changes; for example, when a person's ears pop during a plane ride because the atmospheric pressure lessens with altitude. The atmosphere is thickest (densest) at the planet's surface and gradually becomes thinner (less and less dense) with height above the planet's surface. There is no clear break between the atmosphere and space: the atmosphere just gets thinner and thinner and less and less detectable. Therefore, atmospheric pressure is greatest at the planet's surface and becomes less and less as the height above the planet increases. When the decreasing density of the atmosphere and gravity are taken into consideration, it turns out that atmospheric pressure decreases exponentially with altitude according to the following equation:

$$p(z) = P_o e^{-\alpha z},$$

where $p(z)$ is the atmospheric pressure at some height above the surface z, p_o is the pressure at the surface of the planet, and α is a number that is constant for each planet, and is calculated as follows:

$$\alpha = \frac{g\rho_0}{p_0},$$

where g is the gravitational acceleration of that planet, and ρ_o is the density of the atmosphere at the planet's surface.

Just as pressure diminishes in the atmosphere from the surface of a planet up into space, pressure inside the planet increases with depth. Pressure inside a planet can be approximated simply as the product of the weight of the column of solid material above the point in question and the gravitational acceleration of the planet. In other words, the pressure P an observer would feel if he or she were inside the planet is caused by the weight of the material over the observer's head (approximated as ρh, with h the depth you are beneath the surface and ρ the density of the material between the observer and the surface) being pulled toward the center of the planet by its gravity g:

$$P = \rho g h .$$

The deeper into the planet, the higher the pressure.

to the body, by the transformation of the kinetic energy of the planetesimals into heat in the larger body. Models of the accretion of Jupiter indicate that its original core temperature may have been as high as 180,000°F (100,000°C). To understand kinetic energy, start with momentum, p, which is defined as the product of a body's mass m and its velocity v:

$$p = mv .$$

Sir Isaac Newton called momentum "quality of movement." The greater the mass of the object, the greater its momentum, and likewise, the greater its velocity, the greater its momentum. A change in momentum creates a force, such as a person feels when something bumps into her. The object that bumps into her experiences a change in momentum because it has suddenly slowed down, and what she experiences is a force. The reason she feels more force when someone tosses a full soda to her than when they toss an empty soda can to her is that the full can has a greater mass, and therefore momentum, than the empty can, and when it hits her it loses all its momentum, transferring to her a greater force.

How does this relate to heating by accretion? Those incoming planetesimals have momentum due to their mass and velocity, and when they crash into the larger body, their momentum is converted into energy, in this case, heat. The energy of the body, created by its mass and velocity, is called its kinetic energy. Kinetic energy is the total effect of changing momentum of a body, in this case, as its velocity slows down to zero. Kinetic energy is expressed in terms of mass m and velocity v:

$$K = \frac{1}{2}mv^2.$$

Students of calculus might note that kinetic energy is the integral of momentum with respect to velocity:

$$K = \int mv\,dv = \frac{1}{2}mv^2.$$

The kinetic energy is converted from mass and velocity into heat energy when it strikes the growing body. This energy, and therefore heat, is considerable. The huge heat of accretion has been slowly radiating out of the planet ever since. Jupiter still produces twice the heat that it receives from the Sun; therefore Jupiter's energy budget is not dominated by the Sun, as Earth's energy budget is. In fact, Jupiter was probably so hot from the heat of accretion when it first formed that it would have glowed in visible light, like a red-hot piece of metal does, and radiated so much heat that it blew the atmospheric gases away from its moons.

The center of the core of Jupiter is thought to be about 41,400°F (23,000°C). Measurements of heat flow at Jupiter's surface have been made from both ground-based observations and from *Voyager,* and the heat flowing through Jupiter's surface now supports a core temperature of about this magnitude. Though this means that the core has lost three-quarters of its heat of accretion, it is still hotter than the surface of the Sun. The core is thought to be a solid nucleus of only about 3 percent of Jupiter's mass, with a pressure estimated to be between 30,000,000 to 45,000,000 atm (3,000 and 4,500 GPa). The pressure at the center of the Earth is only about 3,553,200 atm (360 GPa), even though Earth is made of much denser material.

MAGNETIC FIELD

As shown in the photo on page 31, Jupiter's huge size, fast spin, and liquid metallic hydrogen interior make a perfect circumstance for a huge magnetic field, and Jupiter does, in fact, have the largest, strongest magnetic field of all the planets. The planet's fast spin results in currents in the vast inner region of metallic hydrogen that reaches from below about 0.8 Jovian radii down to the core. These currents create a magnetic dynamo, much as currents in the outer, liquid iron core on Earth produce its magnetic dynamo.

Sir Joseph Larmour, an Irish physicist and mathematician, first proposed the hypothesis that the Earth's active magnetic field might be explained by the way the moving fluid iron in Earth's outer core mimics electrical currents, and the fact that

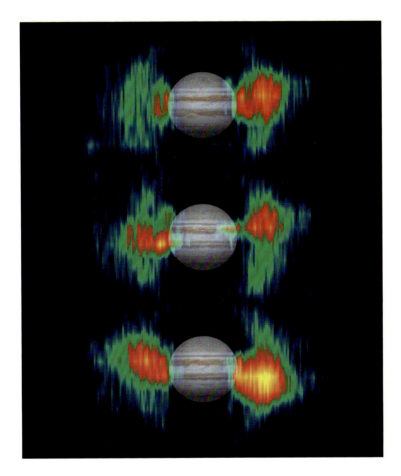

Details in radiation belts close to Jupiter are mapped from measurements that NASA's Cassini spacecraft made of radio emissions from high-energy electrons moving within the magnetic field at nearly the speed of light. The views, with an image of Jupiter superimposed, were taken at three times during the planet's 10-hour rotational period. (NASA/JPL)

every electric current has an associated, enveloping magnetic field. The combination of convective currents and a spinning inner and outer core is called the dynamo effect. If the fluid motion is fast, large, and conductive enough, then a magnetic field can not only be created but also carried and deformed by the moving fluid (this is also what happens in the Sun). The inner core rotates faster than the outer core, pulling field lines into itself and twisting them. Fluid upwelling from the boundary with the inner core is twisted by the Coriolis effect, and in turn twists the magnetic field. In these ways, it is thought that the field lines are made extremely complex. The exact patterns of the field at the core are not known. On Jupiter the flowing currents of metallic hydrogen act as an electrical current: There are free electrons and charged particles in the metallic hydrogen, just like the liquid iron on Earth. Jupiter's large internal solid core makes the patterns of *convection* in the liquid hydrogen more complex, and it creates a magnetic field that is asymmetric and has more than two poles, as discussed below.

Jupiter's magnetic field is tipped by 10 degrees with respect to its rotation axis, similar to the tip of Earth's magnetic field. Jupiter's magnetic field has the opposite polarization than Earth's: Its north magnetic pole is at its south rotational pole. This means that, according to the conventions of physics, magnetic field lines are imagined to run out from the north rotational pole of the planet and into the south rotational pole on Jupiter. Compasses on Jupiter would point to the south pole, rather than the north. The magnetic field on Earth reverses direction periodically, and Earth's north magnetic field has been at its South Pole many times in Earth's history (this is clearly recorded in the rock record). The Earth may be entering a period of magnetic reversal. Jupiter's magnetic field may well be expected to move and reverse over time, as well.

Jupiter's *magnetic moment* is 19,000 times stronger than Earth's. This intensity results in a magnetic field around the planet so huge that if it were visible, from Earth it would appear several times larger than the full Moon. At Jupiter's equator, the intensity of its magnetic field is about 10 times Earth's, and Jupiter's field is two or three times stronger at its poles than at its equator. Jupiter's field is also asymmetric. Though people tend to think of magnetic fields just in terms of *dipoles,* meaning a system with a north and a south pole like the Earth's, there are other, more complex configurations possible for magnetic fields. The next most complex after the dipole is the quadrupole, in which the field has four poles equally spaced around the sphere of the planet. After the quadrupole comes the octupole, which has eight poles. Earth's magnetic field is thought to lose strength in its dipole field and degenerate into quadrupole and octupole fields as it reverses, and then to reform into the reversed dipole field. In the case of Jupiter, stronger quadrupole and octupole fields exist continuously along with the major dipole field, so its field is asymmetrical and its field lines more complex than Earth's.

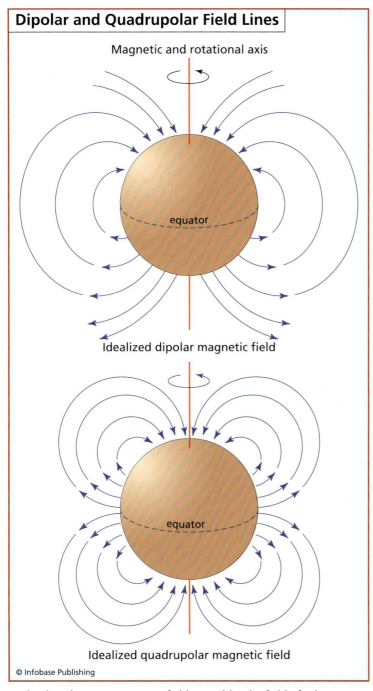

Dipolar and Quadrupolar Field Lines

Magnetic and rotational axis

equator

Idealized dipolar magnetic field

equator

Idealized quadrupolar magnetic field

© Infobase Publishing

A dipolar planetary magnetic field resembles the field of a bar magnet. Jupiter's magnetic field is dominated by a dipole field but has strong components of more complex fields as well.

In the direction toward the Sun, Jupiter's magnetic field extends about 60 Jupiter radii, and has a *bow shock,* a *magnetosheath,* and a *magnetopause,* just as Earth's has. The bow shock is caused by the speedy solar plasma wind striking and piling up against the near side of the magnetic field, like the bow wave of a boat. The bow shock exists at a distance of about 75 Jupiter radii from the planet, and across the bow shock, the solar wind is slowed, compressed, and heated. The shocked solar wind is called the magnetosheath. The inner edge of the magnetosheath is called the magnetopause, inside of which is Jupiter's *magnetosphere,* the volume of its magnetic field. Away from the Sun, the solar wind elongates Jupiter's magnetic field to about 350 Jupiter radii (though this number fluctuates by as much as 100 radii), so long that it almost reaches Saturn's orbit.

Another major difference between Jupiter's and the Earth's magnetic fields is that Jupiter's four innermost moons orbit within Jupiter's magnetic field. Io, the innermost moon, orbits at about 5.9 Jupiter radii and interacts the most with Jupiter's magnetic field. Sulfur dioxide gas (SO_2) flows off of Io at a rate of about a ton per second (this is a similar production rate to that of an active comet). Ionized gases from Io interact with the magnetic field, creating interactions that produce radio emissions detectable from Earth, as described in the next section.

Ganymede, another of the inner satellites, has its own internal magnetic field, and so it is surrounded by its own magnetosphere, buried within Jupiter's.

On Earth, auroras are created by charged particles from the solar wind that are guided into the atmosphere by the Earth's magnetic field, where they bombard atmospheric molecules, which then emit light to release the extra energy given them by the bombardment. Since Jupiter has a massive magnetic field and an atmosphere, it might be expected that it, also, would have auroral displays, though its main atmospheric constituents are hydrogen and helium while on Earth they are nitrogen and oxygen. In 1979 auroras were first observed on Jupiter by a satellite called the *International Ultraviolet Explorer.* Jupiter's auroras are now known to be 1,000 times more energetic than Earth's auroras, and also more complex.

On Jupiter, there is an additional source for auroral emissions beyond the solar wind: The ions shed from Io. Ions from

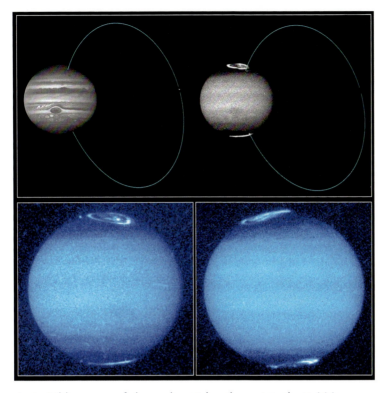

An invisible current of charged particles, drawn into the Hubble Space Telescope *visible light image on the top left, flows from Io along Jupiter's magnetic field lines. The image on the top right, taken in ultraviolet light about 15 minutes later, shows Jupiter's auroral rings. Just outside these rings are spots called footprints, created when the particles from Io fluoresce Jupiter's hydrogen gas. The two ultraviolet images on the bottom show how the auroras change as Jupiter rotates. (John T. Clarke and Gilda E. Ballester [University of Michigan] and John Trauger and Robin Evans [Jet Propulsion Laboratory]/NASA)*

Io fly off onto the magnetic field lines of Jupiter's field, emitting visible light, and sometimes creating a perfect arc away from and back to the planet. The more familiar polar auroras on Jupiter consist of a strong oval around each of the magnetic poles, along with a more diffuse cap of auroral emissions that spread over the high latitudes and extend down past the main oval onto the nightside of the planet. This can be seen in the photo above. While auroras on Earth are green, red, and sometimes blue, due to the characteristic light emissions of ionized oxygen and nitrogen, the hydrogen on Jupiter creates pink auroras.

RADIO NOISE

In 1955, Bernard Burke and Kenneth Franklin, scientists at the Carnegie Institute of Washington, reported bursts of radio noise at 22.2 MHz originating on Jupiter's surface. This is a type of radiation called nonthermal, because it does not originate from the energy that every object with a temperature above absolute zero is radiating at all times; it has some other source for its generation beyond heat. Soon, with additional observations, it was shown that Jupiter emits strong signals at two radio wavelengths, one at tens of centimeters ("decimetric"), and the other at tens of meters ("decametric"). Though since then about 10 radio emission components have been identified, the decimetric and decametric remain the main components. No other planet has so many types of radio emissions.

In 1959, Frank Drake and Hein Hvatum at the National Radio Astronomy Observatory pinpointed the origin of decimetric nonthermal radiation as a toroidal belt around Jupiter, inclined at about 10 degrees to Jupiter's equator. They theorized that the source of this radiation is electrons moving at relativistic speeds while trapped in Jupiter's magnetic field. The tipping of the toroidal region from Jupiter's equator matches the tip of Jupiter's magnetic field, supporting the magnetic field theory. The decimetric radio signals, which are more constant than the decametric, are thought to be caused by electrons orbiting along magnetic field lines and interacting with the motion of the large moon Io. Io has created a toroid of ions around Jupiter, in the shape of its orbit. When the gases that Io emits become ionized, they are attracted along Jupiter's magnetic field lines and rapidly accelerated to 34 miles per second (55 km/sec). When the ions collide or are cooled, they give off visible and radio emissions. The decametric emissions, on the other hand, come in strong, sporadic bursts, and are thought to be the results of magnetic storms, similar in some ways to those that occur on the Sun.

The strength and size of Jupiter's magnetic field create phenomena never seen on Earth: auroral displays caused by ions stripped off a moon orbiting far inside the magnetic field, and pink auroras at its poles as well as elsewhere around its globe where its non-dipolar magnetic field dives back into the planet. The planet's interior experiences such great pressures because of its mass that hydrogen and helium, almost invariably gaseous on Earth, are pressed into liquids, and finally into a metallic form. Under the great pressure of its mass and the heat from its sinking helium rain, along with primordial heat of formation, Jupiter's core is thought to be hotter than the surface of the Sun. In all these ways, except for internal nuclear fusion, Jupiter begins to resemble a tiny star with its own solar system of planets (its numerous moons).

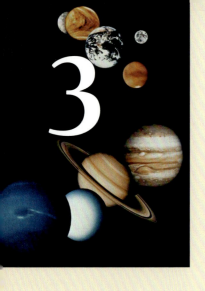

3

Jupiter's Surface Appearance and Conditions

As soon as spectroscopy was developed in the early 20th century, the technique was applied to Jupiter to try to determine what its atmosphere consisted of. (For more, see the sidebar "Remote Sensing" on page 40.) In 1863, Ernest Rutherford, the New Zealand-born geologist and physicist, had used more primitive methods of examining light reflected from Jupiter and discovered features he could not explain. Rupert Wildt, a researcher at Princeton University, was able to use spectroscopy in 1932 to identify Rutherford's anomalies as methane (CH_4) and ammonia (NH_3). These constituents were a bit of a surprise, because hydrogen and helium were known to be the most abundant gas species in the solar system, and so were naturally expected to make up the bulk of Jupiter. Hydrogen was difficult to detect on Jupiter and was not confirmed until the 1960s. Helium was not detected until missions began flying closer to Jupiter. (For more on elements and molecules, see the sidebar "Elements and Isotopes" on page 23.) Since then, techniques for infrared spectroscopy from Earth have been developed, and many more constituents of the Jovian atmosphere have been discovered. Jupiter's atmosphere consists of about 89 percent hydrogen (H_2), 11 percent helium (He), less than a percent of methane, and still smaller amounts of other trace compounds, including phosphine (PH_3), germane (GeH_4 where Ge is germanium, a relatively rare element), water (H_2O), ethane (C_2H_6), and acetylene (C_2H_2).

The highest parts of the atmosphere, called the stratosphere, *thermosphere,* and *exosphere,* have no clouds in them on Jupiter. The boundary at the bottom of the stratosphere is the *tropopause,* the altitude that marks the coldest temperature in the atmosphere. Above the tropopause, the temperature rises, and below the tropopause, the temperature also rises. Above the tropopause, temperature increases due to absorption of solar radiation by high-atmosphere molecules like hydrogen (H_2) and methane (CH_4). Below the tropopause, temperature increases due to heat conducted out of the depths of the planet.

The highest clouds on Jupiter exist just below the tropopause. Jupiter's tropopause has a temperature of about -256°F (-160°C) and a pressure of about 0.6 bar. There are two or possibly three layers of clouds at different heights in the atmosphere. The first is an ammonia (NH_3) cloud layer at around 0.5 bars. Ammonia freezes in the cold temperatures of Jupiter's atmosphere at these pressures, forming the white cirrus clouds that are shaped into zones, ovals, and plumes. The second has its base at 1.3 bars and seems to be made of ammonium hydrosulfide (NH_4SH). The temperature at the tops of the ammonium hydrosulfide clouds is about -58°F (-50°C), and trace elements and compounds of sulfur (S) or phosphorus (P) are thought to give them their sandy or red colors. There is thought to be a thin water (H_2O) cloud below the ammonium hydrosulfide cloud. Jupiter seems to be depleted in water compared to solar abundances, and so it may not have the thick layer of water clouds predicted by thermodynamics to condense at about five bars, where the temperature is about 68°F (20°C). Since the *Galileo* probe entered the atmosphere in a dry, cloud-free area, and the five-bar level is not visible by remote sensing, scientists do not know if a planet-covering layer of water clouds exists. *Galileo* did detect numerous lightning flashes coming from anticyclonic zones in Jupiter's atmosphere. The lightning flashes originated in formations that

(continues on page 48)

REMOTE SENSING

Remote sensing is the name given to a wide variety of techniques that allow observers to make measurements of a place they are physically far from. The most familiar type of remote sensing is the photograph taken by spacecraft or by giant telescopes on Earth. These photos can tell scientists a lot about a planet; by looking at surface topography and coloration photo geologists can locate faults, craters, lava flows, chasms, and other features that indicate the weather, volcanism, and tectonics of the body being studied. There are, however, critical questions about planets and moons that cannot be answered with visible-light photographs, such as the composition and temperature of the surface or atmosphere. Some planets, such as Venus, have clouds covering their faces, and so even photography of the surface is impossible.

For remote sensing of solar system objects, each wavelength of radiation can yield different information. Scientists frequently find it necessary to send detectors into space rather than making measurements from Earth, first because not all types of electromagnetic radiation can pass through the Earth's atmosphere (see figure, opposite), and second, because some electromagnetic emissions must be measured close to their sources, because they are weak, or in order to make detailed maps of the surface being measured.

Spectrometers are instruments that spread light out into spectra, in which the energy being emitted at each wavelength is measured separately. The spectrum often ends up looking like a bar graph, in which the height of each bar shows how strongly that wavelength is present in the light. These bars are called spectral lines. Each type of atom can only absorb or emit light at certain wavelengths, so the location and spacing of the spectral lines indicate which atoms are present in the object absorbing and emitting the light. In this way, scientists can determine the composition of something simply from the light shining from it.

Below are examples of the uses of a number of types of electromagnetic radiation in remote sensing.

GAMMA RAYS

Gamma rays are a form of electromagnetic radiation; they have the shortest wavelength and highest energy. High-energy radiation such as X-rays and gamma rays are absorbed to a great degree by the Earth's atmosphere, so it is not possible to measure their production by solar system bodies without sending measuring devices into space. These high-energy radiations are created only by high-energy events, such as mat-

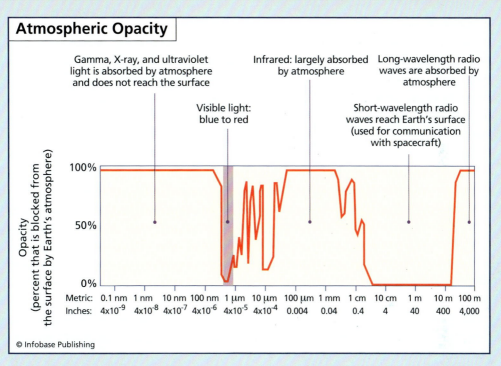

Atmospheric Opacity

Gamma, X-ray, and ultraviolet light is absorbed by atmosphere and does not reach the surface

Infrared: largely absorbed by atmosphere

Long-wavelength radio waves are absorbed by atmosphere

Visible light: blue to red

Short-wavelength radio waves reach Earth's surface (used for communication with spacecraft)

Opacity (percent that is blocked from the surface by Earth's atmosphere)

Metric:	0.1 nm	1 nm	10 nm	100 nm	1 μm	10 μm	100 μm	1 mm	1 cm	10 cm	1 m	10 m	100 m
Inches:	4×10^{-9}	4×10^{-8}	4×10^{-7}	4×10^{-6}	4×10^{-5}	4×10^{-4}	0.004	0.04	0.4	4	40	400	4,000

© Infobase Publishing

The Earth's atmosphere is opaque to many wavelengths of radiation but allows the visible and short radio wavelengths through to the surface.

ter heated to millions of degrees, high-speed collisions, or cosmic explosions. These wavelengths, then, are used to investigate the hottest regions of the Sun. The effects of gamma rays on other solar systems bodies, those without protective atmospheres, can be measured and used to infer compositions. This technique searches for radioactivity induced by the gamma rays.

Though in the solar system gamma rays are produced mainly by the hottest regions of the Sun, they can also be produced by colder bodies through a chain reaction of events, starting with high-energy cosmic rays. Space objects are continuously bombarded with cosmic rays, mostly high-energy protons. These high-energy protons strike the surface materials, such as dust and rocks, causing nuclear reactions in the atoms of the surface material. The reactions produce neutrons, which collide with surrounding

(continues)

(continued)

nuclei. The nuclei become excited by the added energy of neutron impacts, and reemit gamma rays as they return to their original, lower-energy state. The energy of the resultant gamma rays is characteristic of specific nuclear interactions in the surface, so measuring their intensity and wavelength allow a measurement of the abundance of several elements. One of these is hydrogen, which has a prominent gamma-ray emission at 2.223 million electron volts (a measure of the energy of the gamma ray). This can be measured from orbit, as it has been in the Mars Odyssey mission using a gamma-ray spectrometer. The neutrons produced by the cosmic ray interactions discussed earlier start out with high energies, so they are called fast neutrons. As they interact with the nuclei of other atoms, the neutrons begin to slow down, reaching an intermediate range called epithermal neutrons. The slowing-down process is not too efficient because the neutrons bounce off large nuclei without losing much energy (hence speed). However, when neutrons interact with hydrogen nuclei, which are about the same mass as neutrons, they lose considerable energy, becoming thermal, or slow, neutrons. (The thermal neutrons can be captured by other atomic nuclei, which then can emit additional gamma rays.) The more hydrogen there is in the surface, the more thermal neutrons relative to epithermal neutrons. Many neutrons escape from the surface, flying up into space where they can be detected by the neutron detector on Mars Odyssey. The same technique was used to identify hydrogen enrichments, interpreted as water ice, in the polar regions of the Moon.

X-RAYS

When an X-ray strikes an atom, its energy can be transferred to the electrons or biting the atom. This addition of energy to the electrons makes one or more electrons leap from their normal orbital shells around the nucleus of the atom to higher orbital shells, leaving vacant shells at lower energy values. Having vacant, lower-energy orbital shells is an unstable state for an atom, and so in a short period of time the electrons fall back into their original orbital shells, and in the process emit another X-ray. This X-ray has energy equivalent to the difference in energies between the higher and lower orbital shells that the electron moved between. Because each element has a unique set of energy levels between electron orbitals, each element produces X-rays with energies that are characteristic of itself and no other element. This method can be used remotely from a satellite, and it can also be used directly on tiny samples of material placed in a laboratory instrument called an electron microprobe, which measures the composition of the material based on the X-rays the atoms emit when struck with electrons.

VISIBLE AND NEAR-INFRARED

The most commonly seen type of remote sensing is, of course, visible light photography. Even visible light, when measured and analyzed according to wavelength and intensity, can be used to learn more about the body reflecting it.

Visible and near-infrared reflectance spectroscopy can help identify minerals that are crystals made of many elements, while other types of spectrometry identify individual types of atoms. When light shines on a mineral, some wavelengths are absorbed by the mineral, while other wavelengths are reflected back or transmitted through the mineral. This is why things have color to the eye: Eyes see and brains decode the wavelengths, or colors, that are not absorbed. The wavelengths of light that are absorbed are effectively a fingerprint of each mineral, so an analysis of absorbed versus reflected light can be used to identify minerals. This is not commonly used in laboratories to identify minerals, but it is used in remote sensing observations of planets.

The primary association of infrared radiation is heat, also called thermal radiation. Any material made of atoms and molecules at a temperature above absolute zero produces infrared radiation, which is produced by the motion of its atoms and molecules. At absolute zero, -459.67°F (-273.15°C), all atomic and molecular motion ceases. The higher the temperature, the more they move, and the more infrared radiation they produce. Therefore, even extremely cold objects, like the surface of Pluto, emit infrared radiation. Hot objects, like metal heated by a welder's torch, emit radiation in the visible spectrum as well as in the infrared.

In 1879 Josef Stefan, an Austrian scientist, deduced the relation between temperature and infrared emissions from empirical measurements. In 1884 his student Ludwig Boltzmann derived the same law from thermodynamic theory. The relation gives the total energy emitted by an object (E) in terms of its absolute temperature in Kelvin (T), and a constant called the Stefan-Boltzmann constant (equal to 5.670400×10^{-8} W m^{-2} K^{-4}, and denoted with the Greek letter sigma, σ):

$$E = \sigma T^4$$

This total energy E is spread out at various wavelengths of radiation, but the energy peaks at a wavelength characteristic of the temperature of the body emitting the energy. The relation between wavelength and total energy, Planck's Law, allows scientists to determine the temperature of a body by measuring the energy it emits.

(continues)

(continued)

The hotter the body, the more energy it emits at shorter wavelengths. The surface temperature of the Sun is 9,900°F (5,500°C), and its Planck curve peaks in the visible wavelength range. For bodies cooler than the Sun, the peak of the Planck curve shifts to longer wavelengths, until a temperature is reached such that very little radiant energy is emitted in the visible range.

Humans radiate most strongly at an infrared wavelength of 10 microns (*micron* is another word for micrometer, one millionth of a meter). This infrared radiation is what makes night vision goggles possible: Humans are usually at a different temperature than their surroundings, and so their shapes can be seen in the infrared.

Only a few narrow bands of infrared light make it through the Earth's atmosphere without being absorbed, and can be measured by devices on Earth. To measure infrared emissions, the detectors themselves must be cooled to very low temperatures, or their own infrared emissions will swamp those they are trying to measure from elsewhere.

In thermal emission spectroscopy, a technique for remote sensing, the detector takes photos using infrared wavelengths and records how much of the light at each wavelength the material reflects from its surface. This technique can identify minerals and also estimate some physical properties, such as grain size. Minerals at temperatures above absolute zero emit radiation in the infrared, with characteristic peaks and valleys on plots of emission intensity versus wavelength. Though overall emission intensity is determined by temperature, the relationships between wavelength and emission intensity are determined by composition. The imager for *Mars Pathfinder,* a camera of this type, went to Mars in July 1997 to take measurements of light reflecting off the surfaces of Martian rocks (called reflectance spectra), and this data was used to infer what minerals the rocks contain.

When imaging in the optical or near-infrared wavelengths, the image gains information about only the upper microns of the surface. The thermal infrared gives information about the upper few centimeters, but to get information about deeper materials, even longer wavelengths must be used.

RADIO WAVES

Radio waves from outside the Earth do reach through the atmosphere and can be detected both day and night, cloudy or clear, from Earth-based observatories using huge metal dishes. In this way, astronomers observe the universe as it appears in radio waves. Images like photographs can be made from any wavelength of radiation coming from a body: Bright regions on the image can correspond to more intense radiation,

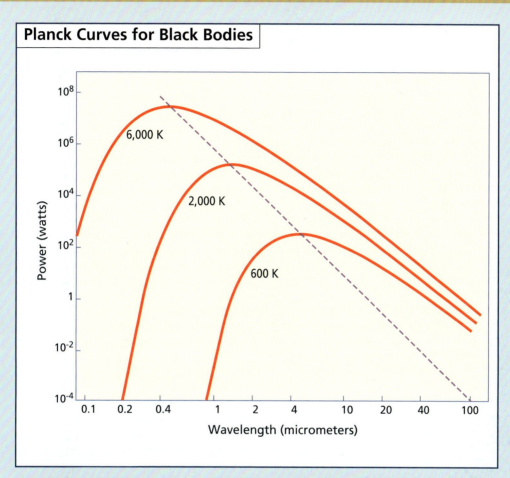

Planck Curves for Black Bodies

The infrared radiation emitted by a body allows its temperature to be determined by remote sensing; the curves showing the relationship between infrared and temperature are known as Planck curves.

and dark parts, to less intense regions. It is as if observers are looking at the object through eyes that "see" in the radio, or ultraviolet, or any other wavelength, rather than just visible. Because of a lingering feeling that humankind still observes the universe exclusively through our own eyes and ears, scientists still often refer to "seeing" a body in visible wavelengths and to "listening" to it in radio wavelengths.

(continues)

(continued)

Radio waves can also be used to examine planets' surfaces, using the technique called radar (radio detection and ranging). Radar measures the strength and round-trip time of microwave or radio waves that are emitted by a radar antenna and bounced off a distant surface or object, thereby gaining information about the material of the target. The radar antenna alternately transmits and receives pulses at particular wavelengths (in the range 1 cm to 1 m) and polarizations (waves polarized in a single vertical or horizontal plane). For an imaging radar system, about 1,500 high-power pulses per second are transmitted toward the target or imaging area. At the Earth's surface, the energy in the radar pulse is scattered in all directions, with some reflected back toward the antenna. This backscatter returns to the radar as a weaker radar echo and is received by the antenna in a specific polarization (horizontal or vertical, not necessarily the same as the transmitted pulse). Given that the radar pulse travels at the speed of light, the measured time for the round trip of a particular pulse can be used to calculate the distance to the target.

Radar can be used to examine the composition, size, shape, and surface roughness of the target. The antenna measures the ratio of horizontally polarized radio waves sent to the surface to the horizontally polarized waves reflected back, and the same for vertically polarized waves. The difference between these ratios helps to measure the roughness of the surface. The composition of the target helps determine the amount of energy that is returned to the antenna: Ice is "low loss" to radar, in other words, the radio waves pass straight through it the way light passes through window glass. Water, on the other hand, is reflective. Therefore, by measuring the intensity of the returned signal and its polarization, information about the composition and roughness of the surface can be obtained. Radar can even penetrate surfaces and give information about material deeper in the target: By using wavelengths of 3, 12.6, and 70 centimeters, scientists can examine the Moon's surface to a depth of 32 feet (10 m), at a resolution of 330 to 985 feet (100 to 300 m), from the Earth-based U.S. National Astronomy and Ionosphere Center's Arecibo Observatory!

(continued from page 39)

seem to be immense thunderstorms, and they are at the level in Jupiter where water clouds are expected to exist. Water, because of its ability to carry a slight electric charge, is efficient at creating lightning. Shown in the photo on page 48, these lightning storms may be indirect evidence that water clouds do exist on Jupiter.

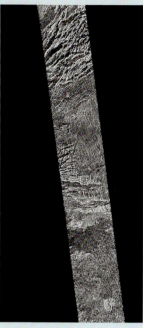

The far greater resolution obtained by the Magellan craft (right) shows the relative disadvantage of taking images of Venus from the Earth (left) using the Arecibo observatory. (NASA/Magellan/JPL)

Venus is imaged almost exclusively in radar because of its dense, complete, permanent cloud cover. Radar images of Venus have been taken by several spacecraft and can also be taken from Arecibo Observatory on Earth. The image below makes a comparison between the resolution possible from Earth using Arecibo (left), and the resolution from the *Magellan* spacecraft (right). Arecibo's image is 560 miles (900 km) across and has a resolution of 1.9 miles (3 km). The *Magellan* image corresponds to the small white rectangle in the Arecibo image, 12 × 94 miles (20 × 120 km) in area. Magellan's resolution is a mere 400 feet (120 m) per pixel.

In the deeper tropopause there seem to be enrichments in carbon (C), sulfur (S), phosphorus (P), and nitrogen (N), to the point that they are more enriched there than in the Sun (the Sun is a convenient comparison because it represents the average bulk composition of the solar system, and so saying that carbon is more enriched than in the Sun means that carbon is more enriched than on average in the solar system).

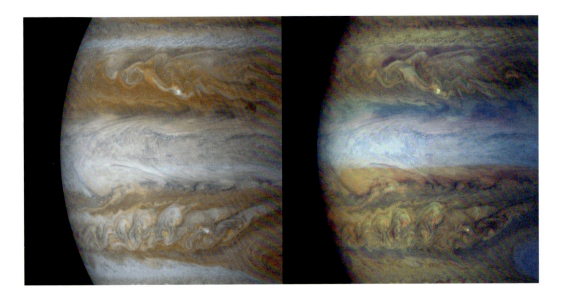

At left is a Cassini natural-color image of Jupiter in visible light. The right frame is composed of three images taken in different parts of the electromagnetic spectrum, yielding colors that show cloud height. Red regions are deepwater clouds, bright blue regions are high haze (like the blue covering the Great Red Spot), small, intensely white spots are energetic lightning storms, and the darkest blue regions are deep hot spots. (NASA/ JPL/University of Arizona)

The *Galileo* probe also measured enrichments in the heavy elements krypton (K), xenon (Xe), and argon (Ar). All of these enrichments are on the order of two to three times the solar values.

The enrichment in nitrogen, and some of the other elements, is a problem in terms of planetary formation. Nitrogen is a *volatile* element and so should be gaseous and exist at solar values at all but the coldest temperatures. To condense a higher proportion of nitrogen than is found in the Sun, the temperature has to be about -400°F (-240°C). This is much colder than space in the vicinity of Jupiter's orbit, where the temperature is about -166°F (-110°C), and, in fact, much colder than temperatures past Uranus and Neptune, in the vicinity of the Oort cloud comets, where space is still about -355°F (-215°C). There are at least two theories that have been presented to explain this quandary. One is that the planetesimals that formed Jupiter came from the very outermost reaches of the solar system, where temperatures were cold enough to condense nitrogen. A second theory is that all of Jupiter formed much farther out from the Sun, in those cold regions. This theory is considered more seriously now that giant, super-Jupiter-sized gas planets have been found orbiting very closely around their stars in other solar systems elsewhere in this galaxy. A huge mass like

Jupiter, upon losing only small amounts of its velocity, could fall into a substantially lower orbit than its original.

The visible surface of Jupiter is about 625 miles (1,000 km) deep into its atmosphere because the vast majority of that depth is clear. (For more, see the sidebar "Optical Depth" on page 53.) The cloud layer is very thin, and in the large scale consists of bands parallel to the planet's equator, covering the entire planet. The bands on the planet are stable enough that they have been assigned names, as listed in the table on page 50. These bands are dark belts and light zones created by strong east-west winds in Jupiter's upper atmosphere. The range of colors seen in the clouds are not caused by the ammonia or water that make up most of the cloud composition but by minute amounts of sulfur and phosphorus compounds so colorful that they serve to color the clouds of ammonia they are in. The light zones are regions of upwelling air, forming

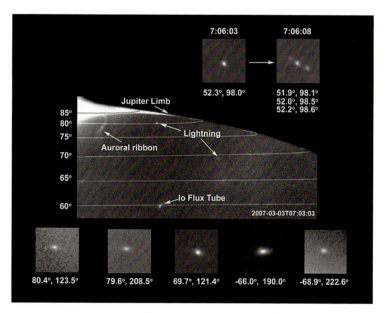

New Horizons has taken the first images of polar lightning on Jupiter. Polar lightning shows that convection in the atmosphere is global and driven by an internal heat source. Its power is consistent with previous low-latitude lightning measurements, equivalent to extremely bright terrestrial "super bolts." (NASA/Johns Hopkins University Applied Physics Laboratory/Southwest Research Institute)

NAMES FOR JUPITER'S BELTS AND ZONES

Definition	Abbreviation	Approximate latitude
north polar region	NPR	47 to 90 degrees
north-north temperate belt	NNTB	43 degrees
north temperate zone	NTZ	35 degrees
north temperate belt	NTB	23 to 35 degrees
north tropical zone	NTRZ	15 to 20 degrees
north equatorial belt	NEB	10 to 15 degrees
equatorial zone (north)	EZn	3 degrees
equatorial belt	EB	0 degrees
equatorial zone (south)	EZs	-3 degrees
southern equatorial belt	SEB	-10 to -22 degrees
Great Red Spot	GRS	-22 degrees
south tropical zone	STRZ	-25 degrees
south temperate belt	STB	-29 degrees
south temperate zone	STZ	-37 degrees
white oval south	WOS	-35 to -37 degrees
south-south temperate belt	SSTB	-41 degrees
south polar region	SPR	-45 to -90 degrees

ammonia cirrus clouds. The dark belts are regions where the cooler atmosphere moves downward. Because there are no ammonia clouds, instruments can see deeper into the atmosphere, to where the clouds are highly colored. The speed of the main equatorial jet is about 190 miles per hour (MPH [300 km/h]), but local winds in cloud tops are much higher.

The striking zonal patterns and very high winds on Jupiter have been a mystery for planetary scientists (zonal patterns circle a planet parallel to the equator, while sectoral patterns are in vertical slices from pole to pole, like orange sections). Some think that the fast rotation is only in a thin upper atmosphere, driven by the very slight solar heat input. Others think that the strong rotation is generated deep in the planet by convection and transferred to the surface. New mathematical modeling efforts have reproduced the circulation patterns on Jupiter, Saturn, and Neptune. The researchers, from the University of South Florida and Ben-Gurion University of the Negev, found that when there was little internal friction in the fluid being studied, with fast rotation of the planet and a low input of heat from the Sun, the planetary atmosphere organized itself into the banded patterns seen on these planets.

In Jupiter's lower atmosphere, the troposphere, there is strong convection. Convection is mixing due to temperature or density differences in a liquid or gas. One example of convection is boiling oatmeal in a pot: Heat put in at the bottom of the pot causes the water and oatmeal at the bottom to expand. Almost every material expands when it is heated, and since it retains its original mass, it becomes less dense. Lowering the density of the material at the bottom makes it buoyant, and it rises to the top. Cooler material from the top sinks to the bottom to take its place, and the cycle continues. Gentle convection can create "cells," regular patterns of circulation with material moving up and down. More violent convection can become turbulent, with currents moving on many length scales. In general, though, convection is caused by a density difference between adjacent pieces of material; the light piece will rise, and the denser piece will sink, initiating a cycle. The differences in density can be caused by temperature, as in the boiling oatmeal case, or the differences in density can be caused by differing compositions.

The troposphere is defined as the lowest part of the atmosphere, where temperature falls with increasing altitude (see figure on page 52). This means that colder, denser gases from the upper troposphere will be continuously falling down to displace the more buoyant and warmer lower atmosphere. The troposphere is thus always mixing, and its currents are wind, creating weather.

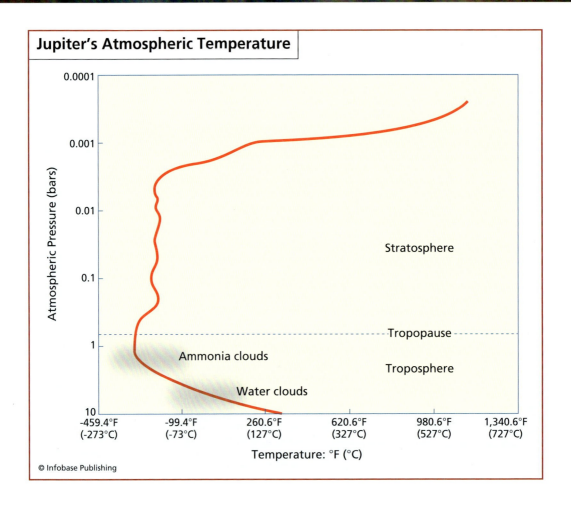

Jupiter's Atmospheric Temperature

Temperatures in Jupiter's atmosphere have been measured remotely and by probes.

The difference in temperature between the polar regions and the equatorial regions drive winds, and therefore weather, just as it does on Earth. There seem to be large zonal winds that roughly correlate with the banded structure. They alternate in direction between prograde and retrograde, and are so strong that they alter the rotation of the planet: They average 350 MPH (150 m/sec) in the cloud tops! The equatorial region of Jupiter makes one rotation in nine hours, 50 minutes, and 30 seconds, while the midlatitudes rotate in nine hours, 55 minutes, and 40 seconds. These strong zonal winds appear to be fairly constant in their locations and strengths, since measurements made by *Voyager* and *Galileo* agree almost perfectly.

OPTICAL DEPTH

Optical depth (usually denoted τ) gives a measure of how opaque a medium is to radiation passing through it. In the sense of planetary atmospheres, optical depth measures the degree to which atmospheric particles interact with light: Values of τ less than one mean very little sunlight is scattered by atmospheric particles or has its energy absorbed by them, and so light passes through the atmosphere to the planetary surface. Values of τ greater than one mean that much of the sunlight that strikes the planet's outer atmosphere is either absorbed or scattered by the atmosphere, and so does not reach the planet's surface. Values of τ greater than one for planets other than Earth also mean that it is hard for observers to see that planet's surface using an optical telescope.

Optical depth measurements use the variable z, meaning height above the planet's surface into its atmosphere. In the planetary sciences, τ is measured downward from the top of the atmosphere, and so τ increases as z decreases, so that at the planet's surface, τ is at its maximum, and z is zero. Each increment of τ is written as $d\tau$. This is differential notation, used in calculus, meaning an infinitesimal change in τ. The equation for optical depth also uses the variable κ (the Greek letter kappa) to stand for the opacity of the atmosphere, meaning the degree of light that can pass by the particular elemental makeup of the atmosphere. The Greek letter rho (ρ) stands for the density of the atmosphere, and dz, for infinitesimal change in z, height above the planet's surface.

$$d\tau = -\kappa\rho dz$$

Mathematical equations can be read just like English sentences. This one says, "Each tiny change in optical depth ($d\tau$) can be calculated by multiplying its tiny change in height (dz) by the density of the atmosphere and its opacity, and then changing the sign of the result" (this sign change is just another way to say that optical depth τ increases as z decreases; they are opposite in sign).

To measure the optical depth of the entire atmosphere, this equation can be used on each tiny increment of height (z) and the results summed (or calculus can be used to integrate the equation, creating a new equation that does all the summation in one step). Optical depth also helps explain why the Sun looks red at sunrise and sunset but white in the middle of the day. At sunrise and sunset the light from the Sun is passing horizontally through the atmosphere, and thus has the greatest distance to travel through the

(continues)

(continued)

atmosphere to reach an observer's eyes. At midday the light from the Sun passes more or less straight from the top to the bottom of the atmosphere, which is a much shorter path through the atmosphere (and let us remember here that no one should ever look straight at the Sun, since the intensity of the light may damage their eyes).

Sunlight in the optical range consists of red, orange, yellow, green, blue, indigo, and violet light, in order from longest wavelength to shortest. (For more information and explanations, see appendix 2, "Light, Wavelength, and Radiation.") Light is scattered when it strikes something larger than itself, like a piece of dust, a huge molecule, or a drop of water, no matter how tiny, and bounces off in another direction. Violet light is the type most likely to be scattered in different directions as it passes through the atmosphere because of its short wavelength, thereby being shot away from the observer's line of sight and maybe even back into space. Red light is the least likely to be scattered, and therefore the most likely to pass through the longest distances of atmosphere on Earth and reach the observer's eye. This is why at sunset and sunrise the Sun appears red: Red light is the color most able to pass through the atmosphere and be seen. The more dust and water in the atmosphere, the more scattering occurs, so the more blue light is scattered away and the more red light comes through, the redder the Sun and sunset or sunrise appear.

The probe from *Galileo* also measured high winds as deep as the 22-bar level. The probe corroborated the measurements of 270 MPH (120 m/sec) at the one-bar cloud level, and found that winds increase in speed to 380 MPH (170 m/sec) at five bars, and remain constant below five bars to the probe's final measuring depth at 22 bars. Finding stronger, constant winds at deeper depths is support for the theory that Jupiter's winds and banded structure are driven by internal convection and heat loss from the planet, rather than from solar input at the surface.

In addition to the strong and constant zonal winds, the *Coriolis force* (the effect of movement on a rotating sphere; movements in the northern hemisphere curve to the right, while movements in the southern hemisphere curve to the left) creates winds of opposite directions north and south of each zone in the Jovian atmosphere, creating many eddies. There are several kinds of oval features on Jupiter's surface that correspond to cyclonic and anticyclonic eddies in Jupiter's atmosphere (*anticyclone* means an area of higher atmospheric pressure than its surroundings, and *cyclone* means an area of lower pressure; each creates vertical winds with circular wind patterns around them). The most prominent is the Great Red Spot, discussed in the section below.

In the North Equatorial Belt there is another kind of strong convective cell. These are called hot spots, because their peculiar atmospheric clarity allows more direct transmission of the great heat of Jupiter's interior to its surface. Despite the heat transmission upward, scientists think that hot spots are associated with strong downwelling winds. The downwelling winds are thought to be responsible for removing clouds from the hot spots by pushing the cloud layers to areas of higher temperature, where the cloud droplets vaporize back into the atmosphere. The hot spots' lack of clouds makes them dark in visible light, since it is Jupiter's clouds that best reflect visible light and thus shine to Earth-based observers. The hot spots are far more visible in the infrared, naturally, because of their anomalously high temperatures, and scientists on Earth have observed them since the first use of infrared astronomy. (They were first reported by the California Institute of Technology scientist James A. Westphal in 1969.) Scientists inadvertently learned more about these North Equatorial Belt hot spots when the probe from *Galileo* flew straight into a hot spot, though that particular path had not been planned. *Galileo*'s probe's measurements, then, are not necessarily characteristic of the rest of the planet.

If downwelling winds vaporize clouds, upwelling winds should produce even denser clouds, by moving warm gas to colder temperatures, where methane, water, and other volatile compounds will condense into droplets and form clouds. (This is the same process that creates drops of condensation on the outside of a cold glass of water: The cold glass conductively cools the surrounding air, and water that had been warm enough to form a vapor loses energy and condenses into a liquid.) Large, bright clouds in Jupiter's equatorial regions are thought to be the sites of upwellings. As might be expected in symmetrical convective cells, the clear downwellings and cloudy upwellings appear to form in pairs, and often are regularly spaced along the equator. Though the number of pairs changes over time, usually there are between nine and 13 pairs around Jupiter's equator.

CHANGES IN JUPITER'S SURFACE WITHIN HUMAN HISTORY: STORMS

As instruments improve and finer details on other planets can be seen, scientists are more often able to watch changes on the surfaces of other planets. The Moon's surface is largely unchanging, since it is no longer volcanically active, it has no atmosphere or large-scale weather, and there have been no recent large impacts onto its surface. Since the Moon often dominates mankind's images of other planets, it is tempting to think of other planets as cold, unchanging places in general. Planets like Jupiter and Saturn, though, have large heat outputs from their interiors and giant, changing magnetic fields, along with active gaseous atmospheres, and so there their surfaces change in ways that can be seen over time by mankind.

The Great Red Spot is a huge anticyclonic storm in Jupiter's southern hemisphere, around 20 degrees south latitude. Giovanni Cassini, the Italian astronomer, may have seen the Great Red Spot on Jupiter first around 1655, or Robert Hooke, the English experimentalist, may have seen it first in 1664. Though telescopes had only developed in the previous few decades to the point that the Great Red Spot could be resolved, there is a school of thought that the Great Red Spot itself first developed on Jupiter in the early 1600s. Early descriptions of Jupiter's cloud systems reported no spot, and it is a very prominent feature, so the reasoning goes that if it had been there in the earliest years of telescopic observation (beginning in 1608), it would have been commented upon. Some historians of science also think that the first convincing report of the Great Red Spot was in 1831 by the scientist Heinrich Schwabe, and that the feature that Hooke described in 1664 was a different storm altogether. The image of the Great Red Spot shown here was taken by the Voyager 2 mission and also shows a white oval, directly beneath the Great Red Spot.

Though the Great Red Spot has remained active throughout at least the last 170 years, it does change visibly through time. The spot did not become truly prominent until the 1880s, which is when it developed its deep red color. Since then the color has continued to fluctuate, varying from deep red to pale salmon or buff, or even disappearing completely, leaving what is then called the Red Spot hollow. The Great Red Spot is now about 15,000 miles (24,000 km) in the east-west dimension × 8,750 miles (14,000 km) in the north-south dimension. Though even the current Great Red Spot is far larger than the planet Earth, which has a diameter of about 8,125 miles (13,000 km), the spot has been as large as 25,000 miles (40,000 km) in its east-west dimension. In addition to changing size, the spot wanders in position. Over a regular 90-day period, the Spot moves 1,250 miles (2,000 km) north and then 1,250 miles (2,000 km) south of its average latitude. The Great Red Spot also moves in longitude, and over the last hundred years has completed about three circuits back and forth around the planet. The latest observations from *Galileo* showed that the

Jupiter's Great Red Spot is an immense anticyclonic storm that fluctuates in size, color, and position over time. (NASA/JPL/ Voyager 2)

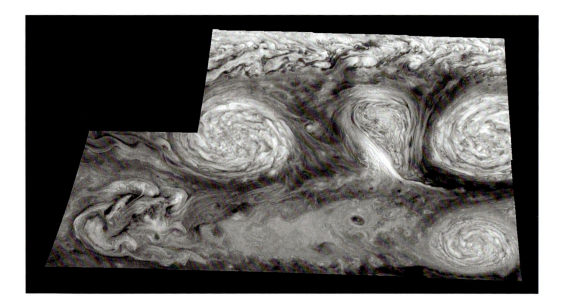

White Ovals are anticyclones smaller and more transient than the Great Red Spot. (NASA/JPL/ Galileo)

interior of the Great Red Spot is rotating more slowly than its edges, and that its very center may even have an area of circulation in the opposite sense from the rest of the spot. Throughout most of the spot, winds cycle counterclockwise, and at its edges, take about six Earth days to make one complete lap around the spot. Though wind speed and size show that this is a huge storm, there are still no satisfactory theories about its cause and duration.

Aside from the Great Red Spot, smaller-scale cyclones and anticyclones develop and fade away rapidly, on timescales of a few months. These tend to appear as white spots, where clouds are lacking. In 1938 in Jupiter's South Temperate Belt, three large anticyclonic ovals were first observed and named the White Oval Spots. In the image above, the leftmost of the White Ovals is 5,600 miles (9,000 km) in breadth. They move relative to each other as the belts move and over time have merged. In both 1998 and 2000, two of the White Oval Spots merged into one. Since 2000, there has been only one White Oval Spot.

The remaining White Oval, known as Oval BA, passes the Great Red Spot about every two years. Scientists from the NASA Goddard Space Flight Center think the two storms might also merge, spelling the end of the almost century-long history of the White Ovals, but Oval BA has so far avoided joining the Great Red Spot. Oval BA, however, is gaining in strength: Its wind speeds have increased along with its size, and in early 2008 a fundamental change in Oval BA indeed ended the history of White Ovals. Oval BA turned red. Some NASA scientists are now calling it the Red Spot, Jr.

Philip Marcus, a professor of physics at the University of California at Berkeley, has conducted computer models that predict global climate changes on Jupiter. Vortices like the White Spots transfer heat from Jupiter's equatorial regions to its poles. Eventually, the equator will heat by as much as 18°F (10°C) and its poles will cool by a similar amount. The extreme heat gradient will force vortices to form again and transfer heat to the poles.

CHANGES IN JUPITER'S SURFACE WITHIN HUMAN HISTORY: COMET SHOEMAKER-LEVY

A major event in the histories of both Jupiter and mankind occurred in July 1994 when the comet Shoemaker-Levy collided with Jupiter and was witnessed by both ground-based observatories and the then-nearby *Galileo* spacecraft. The comet was first discovered in March 1993 by the immensely influential husband-and-wife scientific team of Eugene and Carolyn Shoemaker and their colleague David Levy, using the 1.3-foot (0.4-m) Schmidt telescope on Mount Palomar in California. The comet had already broken into a chain of about 20 fragments, as can be seen in the image from the *Galileo* spacecraft on page 61. As its orbit was better defined, it was identified as a Jupiter-family planet whose orbit had been disrupted by tidal forces from Jupiter in the previous July, and that it would in fact crash into the planet the following July. From modeling efforts of the Jupiter system, it has been estimated that a giant impact of this type should only happen once every few hundred years, and in every way its observation was a great opportunity for the advancement of the understanding of giant impacts and the Jupiter system. (It is possible that such an event had been seen by Giovanni Cassini in 1690 at the Paris Observatory, when he reported and made drawings remarkably similar to the images of the Shoemaker-Levy impacts.)

Observers prepared for the impact, taking images and making measurements at wavelengths from the infrared to the radio, from at least 30 professional ground-based observatories, and from *Galileo,* the *Hubble Space Telescope,* the *International Ultraviolet Explorer,* and the X-ray satellite *ROSAT.* There was a lot of disagreement about what to expect from the impacts. Some scientists predicted that the impacts would be virtually invisible from Earth because they would produce so little energy. Others predicted that the icy comet fragments would burn up like meteors in Jupiter's atmosphere, leaving nothing to make an impact. As it turned out, the impacts were far more exciting than most people had dared hope: The visible light flash from each impact was so bright that it could be seen clearly even with a tiny backyard two-inch refracting telescope at a magnification of 75 times.

The various fragments entered Jupiter's atmosphere between July 16 and 22, 1994, at about 44 degrees south latitude. The first impacts struck the surface just behind the edge of the disk of Jupiter, on the nightside of the planet out of sight from Earth, but they rotated into view after about 10 minutes. By great good timing and fortune, the *Galileo* spacecraft did have a direct view of the impacts. Though Earth-based observers could not see the impacts themselves, they could see flashes of light reflect off Jupiter's inner moons.

The moment of impact of each fragment is badly recorded, and the moment of impact appears to be unimpressive in terms of visible light emissions. About a minute after each impact, though, came the more dramatic explosion phase: A brilliant visible-light flash was created by the explosion of the fragment when it reached high enough temperatures and pressures, and by the rapid rise of the resulting fireball. Two large concentric waves radiated out through the atmosphere from each impact site, moving at 980 to 1,600 feet per second (300 to 500 m/sec), in a mechanism similar to earthquake waves. Starting about six minutes after impact, the splash phase began, when ejected

material began raining back into the atmosphere, causing significant heating. The *ejecta* appears to have been thrown as high as 1,875 miles (3,000 km), and its raining back took about 15 minutes.

The four images above of Jupiter and the bright nightside impact of fragment W of the comet were taken by the *Galileo* spacecraft. The spacecraft was 148 million miles (238 million km) from Jupiter at the time, about 40 degrees from Earth's line of sight to Jupiter, permitting this direct view. The images were taken at intervals of two and one-third seconds, using the green filter in visible light. The first image shows no impact. In the next three images, a point of light appears, brightens so much as to saturate its picture element, and then fades again, seven seconds after the first picture. The location is approximately 44 degrees south as predicted; dark spots visible in the figure above are from previous impacts.

Based on modeling of the radiation emissions of the phases of impact, it has been estimated that the fragments varied in size from about 245 to 985 feet (75 to 300 m) in radius (note that these are much, much smaller than the near-Earth asteroids that are being tracked for possible Earth impact). Each observed impact was given a letter designation starting with A, and there seem to have been 21 impacts. The largest fragment, G, struck Jupiter with an estimated energy

Galileo was positioned to photograph the fragments of comet Shoemaker-Levy as they collided with Jupiter. The impact occurred in a shadowed region of Jupiter in these images, but the energy output saturated the photo elements of the camera. (The impacts were not directly visible from Earth, though their locations of impact rotated into view from Earth shortly after impact.) (NASA/JPL/ Galileo)

equivalent to 6,000,000 megatons of TNT (about 600 times the world's estimated total arsenal). It is not known how deeply the fragments penetrated, though some appeared to explode above the cloud layer, and some below.

The temperatures of the fragments' fireballs reached 18,000°F (10,000°C) immediately, and decreased to about 3,600°F (2,000°C) within 15 seconds. The diameters of the fireballs began at about 10 miles (15 km), and grew to 60 miles (100 km) after 40 seconds. Violent eddies began in the atmosphere around the impact sites. Despite these huge temperatures and sizes, the splash phase, when ejecta reentered the atmosphere, was by far the most consequential for the planet: The atmosphere was significantly heated, and new molecules (H, C, S, O, and N compounds) were formed in the upper atmosphere because of the kinetic heating of the falling ejecta. The atmosphere in the region of the ejecta was heated by hundreds of degrees, and cooling to normal temperatures took a day or two.

Some of the new molecular species caused by the ejecta were still detectable years after the impacts. Each impact site was marked by a brown spot, probably a layer of condensed matter from the fireball and plume, and most were still visible after five months, even in Jupiter's active atmosphere. These effects mirror what scientists have suspected would happen on Earth in the case of a giant impact: The heating of the atmosphere from ejecta would trigger worldwide forest fires, and changing the composition of the atmosphere and injecting dust and ash into it would block solar radiation, perhaps for years. The impacts were such a huge event in planetary science that within two years, over 100 scientific papers about the event had been published.

Since the beginning of space observation through telescopes, in 1608, Jupiter has been a great subject of study. Its immense size and brightness and relative nearness compared to the other gas giant planets have all made it a desirable viewing target. With the development of better and better observational tools, Jupiter has simply become more interesting and informative: Its wildly powerful storms and ever-changing weather patterns are relatively easy to watch and form a kind of window into Jupiter's internal processes and temperature structure. Jupiter's huge gravity field attracted comet Shoemaker-Levy, and its multiple impacts have formed a seemingly infinite source of data for the scientists who study impacts. Scientific papers on that topic are still being published, and Jupiter will no doubt stay a main focus of planetary study.

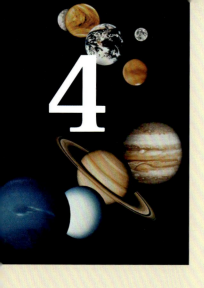

Jupiter's Rings and Moons

Jupiter's many moons, more than any of its other attributes, help it resemble a miniature solar system in its own right, though the relatively recent discovery of rings around Jupiter now suggests an entirely new and complex set of interactions among the orbiting objects around Jupiter and its magnetic and gravity fields.

THE SURPRISE OF JUPITER'S RINGS

Prior to the *Voyager 1* mission in 1979, it was thought that only Saturn and Uranus had rings. Saturn was known as the ringed planet, and its gorgeous rings had become a generalized symbol for planets. Then, when *Voyager 1* visited Jupiter, the scientific world was shocked to see clearly that Jupiter also had rings. Now, scientists have come to think that rings are the norm rather than the exception: Jupiter, Saturn, Uranus, and Neptune all have rings, and some scientists think that even Mars may have very thin and wispy rings.

Jupiter has three rings: a flattened main ring, an inner cloudlike ring called the halo, and a third ring called the gossamer ring. NASA's two *Voyager* spacecraft first revealed that Jupiter has a flattened main ring and an inner halo, both composed of small, dark particles. Jupiter's rings are made of dust rather than ice, making them finer and darker than Saturn's rings. The main ring is 4,400 miles (7,000 km) wide, with a radius of 79,000 miles (126,000 km), but its thickness is a mere half mile (1 km). The halo ring extends from the main ring almost all the way to Jupiter itself, and is the only ring in the solar system to do so. The halo is a faint ring, 12,500 miles (20,000 km) wide. The image below shows the halo (top) and the main ring (bottom) but does not show the gossamer ring.

Only one *Voyager* image showed a hint of the third, faint outer ring. Later *Galileo* data revealed that this third ring, known as the gossamer ring because of its transparency, consists of two rings (see the figure below). One is embedded within the other, and both are composed of microscopic

Jupiter's main and halo rings, discovered in 1979 by Voyager 1, are darker and finer than Saturn's bright rings. (NASA/JPL/Galileo)

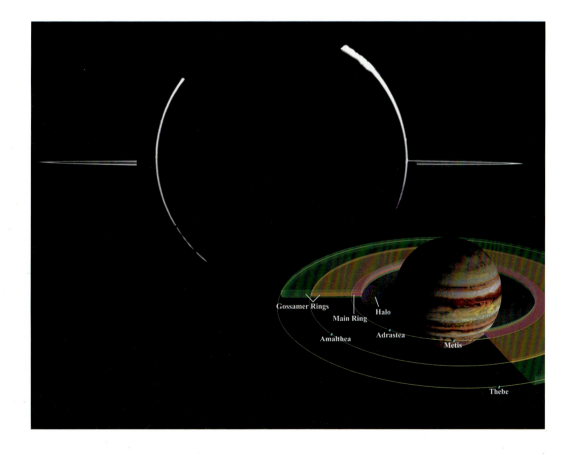

debris. The *Galileo* spacecraft acquired this mosaic of Jupiter's ring system (top) when the spacecraft was in Jupiter's shadow looking back toward the Sun, and a figure showing the positions of the rings is shown at bottom.

Joseph Burns, Maureen Ockert-Bell, Joseph Veverka, and Michael Belton, from Cornell University and the National Optical Astronomy Observatories, have developed theories for how Jupiter's unusual ring system formed. Some of the three dozen *Galileo* images from 1996 and 1997 actually showed the gossamer-bound dust shedding from Amalthea and Thebe, which may be the main sources of the ring material. These images provided one of the most significant discoveries of the entire *Galileo* imaging experiment, because they showed that the gossamer rings are formed in an entirely unexpected way.

The rings contain very tiny particles resembling dark reddish soot. Unlike Saturn's rings, Jupiter's rings show no signs of ice. Scientists believe that dust is kicked off the small moons when meteoroids strike them, and perhaps by magnetic field interactions. Though interplanetary speeds of six to 50 miles per second (10 to 70 km/sec) are normal for asteroids and meteoroids, Jupiter's huge gravity field accelerates the tiny bodies beyond even these huge velocities. Without any atmospheres to slow the meteoroids, they strike the moons so fast that they instantly become buried in the surface of the moon and vaporize and explode from frictional heating and shock.

The impacts of the meteoroids cause clouds of dust particles to be thrown off the moon and into Jupiter orbit (these inner moons are so small that their escape velocity is negligible). As dust particles are blasted off the moons, they enter orbits that are much like those of their source satellites, both in their distance from Jupiter and in their slight tilt relative to Jupiter's equatorial plane. The range of orbits forms a disk shaped like a washer, with a flat outer edge. This explains why, in *Galileo*'s close-up, edge-on view, the ring-tip profiles are rectangular rather than the familiar elliptical arc seen on Saturn's rings and Jupiter's main ring.

The innermost halo ring appears to contain escaped particles from the main ring. The particles are electrically charged and interact with Jupiter's enormous electromagnetic force, causing the cloud of particles to swell into a vast cloud slowly drawn down into the planet.

JUPITER'S MOONS

In 1608, Hans Lippershey, a Dutch eyeglass-maker, attached a lens to each end of a hollow tube and thus created the first telescope. Galileo Galilei, the Italian astronomer born in Pisa in 1564, made his first telescope in 1609 from Lippershey's model. In 1610 Galileo discovered four of Jupiter's moons: Io, Europa, Ganymede, and Callisto. They are still called the Galilean satellites, or Galilean moons. These were the first bodies in the solar system that could clearly be demonstrated not to be orbiting Earth, and in fact their discovery marked the beginning of the end of the wildly incorrect theory that all heavenly objects orbit the Earth.

Galileo himself sent a telescope to Johannes Kepler, the prominent German mathematician and astronomer. When Kepler saw the moons for himself he promptly coined the word *satellite* from a Latin term meaning "hangers-on to a prominent man." Thirteen of Jupiter's moons were known before the Voyager space mission, and since Voyager, teams of scientists, including a team at the University of Hawaii, have been searching for new moons. The Hawaii team discovered 11 new moons in 2001, and more are expected. Thirty-nine moons were known in 2002, and by the beginning of 2004, 62 moons had been identified. The satellites were detected using the world's two largest digital telescope cameras at the Subaru (8.3 meter diameter) and Canada-France-Hawaii (3.6 meter diameter) telescopes atop Mauna Kea in Hawaii.

The Galilean satellites, Io, Europa, Ganymede, and Callisto, make a kind of miniature solar system orbiting Jupiter. (See the photo on page 68.) Their densities fall with distance from Jupiter, just as the densities of the planets fall with distance from the Sun. Their density progression is consistent with the moons having formed at the same time as the planet, as are their circular orbits in the plain of Jupiter's equator. Io is volcanically active and covered with silicate minerals and sulfur, while the

The satellites Io, Europa, Ganymede, and Callisto were discovered by Galileo in 1610 and were the first bodies in the solar system that were obviously orbiting a body other than the Earth, which the solar system was then believed to do. (NASA/JPL/Galileo)

other three Galilean satellites are ice covered. This portrait of the Galilean satellites was constructed out of separate images of each moon, but they are shown at approximately the same scale to compare their sizes. The four moons are shown in order of increasing distance from Jupiter from left to right.

Unlike the Galilean satellites, based on their composition and shape many of Jupiter's moons are probably captured asteroids. (See figure on page 69). Moons that were formed of the material that makes up Jupiter, and at the same time as Jupiter, should have a density consistent with their distance from Jupiter, and they should be round from the heat of formation. Irregular or unusually dense moons are probably captured asteroids, as are moons that orbit in a retrograde sense, opposite to natural moons and the planets' direction around the Sun. All of Jupiter's moons keep one face toward

Jupiter at all times as they rotate and revolve, just as the Moon does with Earth. This is known as *synchronous rotation* and is described in the sidebar "What Are Synchronous Orbits and Synchronous Rotation?" on page 72.

All the moons are within the huge magnetosphere of Jupiter, which deflects the solar wind around them. Thus, the surfaces of the moons have no solar gardening (alteration of the surface by solar wind and micrometeorites), though the inner moons are bombarded by ions streaming along Jupiter's own magnetic field.

There is a wonderful connection between coming to understand the orbits of Jupiter's moons and making early measurements of the speed of light. After Galileo discovered his four moons of Jupiter, he challenged others to make careful measurements of their periods. Tables of their movements were developed in 1665 by Giovanni Borelli, the Italian mathematician, and in 1668 by Giovanni Cassini, the Italian astronomer.

The orbital periods of Jupiter's four moons were found to be 1.769 days, 3.551 days, 7.155 days, and 16.689 days. These seemed to be very constant and predictable, just like all other heavenly motions. Based on these figures, it was possible to predict within minutes the times of eclipses and passages (when the planet moves behind and in front of Jupiter) that would occur in future observations. Previous measurements of the moons' movements had been made when Jupiter was in *opposition,* that is, on the opposite side of the Earth from the Sun, as shown in the figure on page 70. Opposition makes Jupiter brighter and easier to see, since the Sun shines directly upon Jupiter from the point of view of the Earth throughout Earth's night. Now knowing when the moons' passages should occur based on their orbital periods, astronomers began to make observations of the moons at times when measurements were more difficult, such as when Jupiter was nearly in *conjunction* with the Sun. At conjunction, the Sun is nearly between the Earth and Jupiter, and it is only possible to observe Jupiter just after sunset or before daybreak.

The eclipses and passages of Jupiter's moons at conjunction, which had been predicted so precisely when Jupiter was in opposition, were consistently later than their predicted

Opposition and conjunction are the two cases when the Earth, the Sun, and the body in question form a line in space.

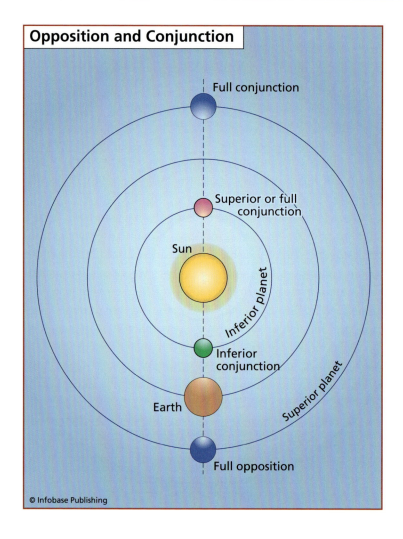

Opposition and Conjunction

Full conjunction

Superior or full conjunction

Sun

Inferior planet

Inferior conjunction

Earth

Superior planet

Full opposition

© Infobase Publishing

occurrence. All four moons seemed to be late by the same amount of time. The early astronomers measured the discrepancy as 22 minutes, but more recent measurements have shown that the moons' movements are slightly less than 17 minutes late. At the subsequent opposition six months later, all the moons were found to be back on their predicted schedule. Cassini recognized these discrepancies, and, at first, he attributed them to a finite speed of light. He wrote in his notes that the discrepancies were due to light taking a certain amount of time to reach us from the moon, and that it takes 10 or 11 minutes for light to travel across the radius of Earth's orbit. This explanation for the data discrepancies is exactly right, but being a scientist of his time, Cassini could not quite believe his own reasoning and tried to find other explanations.

While mulling over this data in 1675 on a visit to Paris, the Danish astronomer Ole Christensen Rømer also thought of an explanation based on the idea that sight is not instantaneous. If light travels at a finite speed, when an observer sees distant things he or she is really seeing how those distant things were at some time in the past. The farther away the observer from an object, the greater the time delay. Applying this hypothesis to the observations of Jupiter's moons, Rømer realized that when Jupiter was in conjunction with the Sun, light from Jupiter and its moons had to travel an extra distance to equal the diameter of the Earth's orbit. Using this distance of two radii of the Earth's orbit (that is, two astronomical units) and the time of delay, Rømer made the first reasonable calculation of the speed of light (about 1AU / 8.5 minutes = 3.0×10^8 m/sec).

The moons of Jupiter are listed in the table on page 75 in order from the moon orbiting nearest Jupiter (Metis) to the farthest moons now known. The Greeks knew Jupiter originally as the god Zeus, and many of the moons of Jupiter are named for characters from Greek and Roman mythology. Metis orbits in an almost perfectly circular path 80,062 miles (128,100 km) from Jupiter, and the farthest moon now known, S/2003 J23, orbits at 15 million miles (24 million km) from the planet. Some of these moons have been well studied and a lot can be written about them, but others are hardly known, often because they are so small. Good reflectance spectra and good images are difficult to obtain from tiny, distant bodies. When poorly known moons are written about here, their compositions and dimensions should be taken as only rough estimates. Only 20 of Jupiter's moons are larger than 10 miles (15 km) in diameter. The other 43 moons now known are often truly tiny objects, less than three miles (2 km) in diameter. If such a thing were possible, a person could walk all the way around their circumference in about three hours. Many of

WHAT ARE SYNCHRONOUS ORBITS AND SYNCHRONOUS ROTATION?

Synchronous rotation can easily be confused with *synchronous orbits.* In a synchronous orbit, the moon orbits always above the same point on the planet it is orbiting (this section uses the terms *moon* and *planet,* but the same principles apply to a planet and the Sun). There is only one orbital radius for each planet that produces a synchronous orbit. Synchronous rotation, on the other hand, is created by the period of the moon's rotation on its axis being the same as the period of the moon's orbit around its planet, and produces a situation where the same face of the moon is always toward its planet. *Tidal locking* causes synchronous rotation.

Gravitational attraction between the moon and its planet produces a tidal force on each of them, stretching each very slightly along the axis oriented toward its partner. In the case of spherical bodies, this causes them to become slightly egg-shaped; the extra stretch is called a tidal bulge. If either of the two bodies is rotating relative to the other, this tidal bulge is not stable. The rotation of the body will cause the long axis to move out of alignment with the other object, and the gravitational force will work to reshape the rotating body. Because of the relative rotation between the bodies, the tidal bulges move around the rotating body to stay in alignment with the gravitational force between the bodies. This is why ocean tides on Earth rise and fall with the rising and setting of its moon, and the same effect occurs to some extent on all rotating orbiting bodies.

these tiny objects orbit at immense distances from the planet, and almost all the outer moonlets orbit in a retrograde sense, that is, in the opposite direction than most moons. This is thought to be a strong indicator that they are trapped asteroids and did not form when Jupiter formed. The very high eccentricities of these orbits compared to the stable inner moons is further evidence for their being trapped asteroids. Trapping asteroids as they go by is a reasonable occurrence to expect from such a huge mass, but keeping trapped asteroids in stable orbits is a physically difficult feat. The orbits tend to degenerate and cause the moonlet eventually to fall into the planet. Many of Jupiter's tiny moonlets therefore may be only

The rotation of the tidal bulge out of alignment with the body that caused it results in a small but significant force acting to slow the relative rotation of the bodies. Since the bulge requires a small amount of time to shift position, the tidal bulge of the moon is always located slightly away from the nearest point to its planet in the direction of the moon's rotation. This bulge is pulled on by the planet's gravity, resulting in a slight force pulling the surface of the moon in the opposite direction of its rotation. The rotation of the satellite slowly decreases (and its orbital momentum simultaneously increases). This is in the case where the moon's rotational period is faster than its *orbital period* around its planet. If the opposite is true, tidal forces increase its rate of rotation and decrease its orbital momentum.

Almost all moons in the solar system are tidally locked with their primaries, since they orbit closely and tidal force strengthens rapidly with decreasing distance. In addition, Mercury is tidally locked with the Sun in a 3:2 *resonance*. Mercury is the only solar system body in a 3:2 resonance with the Sun. For every two times Mercury revolves around the Sun, it rotates on its own axis three times. More subtly, the planet Venus is tidally locked with the planet Earth, so that whenever the two are at their closest approach to each other in their orbits, Venus always has the same face toward Earth (the tidal forces involved in this lock are extremely small). In general any object that orbits another massive object closely for long periods is likely to be tidally locked to it.

temporary visitors to the Jupiter system. The moons with the most known about them are described as follows.

1. Metis

Metis is a small and irregularly shaped moon named after a Titaness who was a consort of Jupiter. Metis and Jupiter's next moon, Adrastea, lie within Jupiter's main ring and may be the source of material for the ring. Metis was discovered in images from *Voyager 1,* but little is known about it. It was not until 1996, when *Galileo* took a fuzzy image of the moon, that Metis was seen as anything more than a point of light. The image on page 78 is one of the best ever taken of Metis. This composite image shows, from left, Jupiter's moons Thebe, Amalthea, and Metis, on approximately the same scale.

2. Adrastea

Adrastea seems to be much like Metis, small and irregularly shaped. Adrastea was named after the daughter of Jupiter and Ananke. Metis was a justice and judgment figure, granting rewards and assigning punishments. Like Metis, Adrastea was first photographed by *Voyager* in the 1970s, but the moon appeared as nothing more than a point of light until *Galileo* arrived in the 1990s. Adrastea was discovered from Voyager mission data by David Jewitt while still a graduate student. Both Metis and Adrastea orbit synchronously and with Jupiter's *Roche limit*. (See the sidebar "Why Are There Rings?" on page 144.) They may be too small to be torn apart by tidal stresses, but their orbits will eventually decay to the point that they will fall into Jupiter.

3. Amalthea

Amalthea is named for a nymph who nursed the baby Jupiter. Though not as tiny as little Metis and Adrastea, Amalthea is a small moon, measuring 169 × 103 × 94 miles (270 × 165 × 150 km). Despite its small size, Amalthea was the very last moon in the solar system discovered with a ground-based optical telescope. Edward Barnard found it in 1892 while observing through the Lick Observatory's 36-inch (91-cm) refracting scope. Barnard became something of a star of astronomy, because this was the first of Jupiter's moons discovered since Galileo found the four largest moons in 1610.

Amalthea orbits at 1.5 Jupiter radii, passing around Jupiter once every 12 hours. Amalthea orbits synchronously, with its long axis pointing in toward Jupiter. Its surface is very old and covered with craters. The largest crater, Pan, is 60 miles (100 km) in diameter, and five miles (8 km) deep. Another large crater, Gaea, is twice as deep as Pan but only 50 miles (80 km) in diameter. Amalthea also has at least two mountains, Mons Lyctas and Mons Ida, which approach 13 miles (20 km) high. Imagine a world only 170 miles (274 km) long at best, not quite

(continues on page 78)

JUPITER'S MOONS

Moon	Radius (miles [km])	Orbital Eccentricity	Year Discovered
Small inner satellites			
1. Metis	14 (22)	0.001	1979
2. Adrastea	5 (8)	0.002	1979
3. Amalthea	85 by 51 by 47 (135 by 82 by 75)	0.003	1892
4. Thebe	34 by 28 (55 by 45)	0.018	1979
Galilean satellites			
5. Io	1,134 (1,822)	0.000	1610
6. Europa	975 (1,561)	0.000	1610
7. Ganymede	1,645 (2,631)	0.001	1610
8. Callisto	1,506 (2,411)	0.007	1610
Themisto prograde group			
9. Themisto	3 (4.5)	0.242	2000
Himalia prograde group			
10. Leda	6 (9)	0.164	1974
11. Himalia	58 (93)	0.162	1904
12. Lysithea	12 (19)	0.112	1938
13. Elara	24 (39)	0.217	1905
14. S/2000 J11	1.3 (2)	0.248	2000
Carpe prograde group			
15. Carpo (S/2003 J20)	1 (1.5)	0.430	2003
Ananke retrograde group			
16. S/2003 J12	0.3 (0.5)	0.376	2003

(continues)

JUPITER'S MOONS *(continued)*

Moon	Radius (miles [km])	Orbital Eccentricity	Year Discovered
Ananke retrograde group			
17. Euporie	5.2 (8.3)	0.144	2001
18. S/2003 J3	0.6 (1)	0.241	2003
19. S/2003 J18	0.6 (1)	0.119	2003
20. Orthosie	5.2 (8.3)	0.281	2001
21. Euanthe	5 (8)	0.232	2001
22. Harpalyke	4.8 (7.6)	0.226	2000
23. Praxidike	4.8 (7.5)	0.230	2000
24. Thyone	5 (8)	0.229	2001
25. S/2003 J16	0.6 (1)	0.270	2003
26. Mneme (S/2003 J21)	0.6 (1)	0.227	2003
27. Iocaste	4.6 (7.3)	0.216	2000
28. Helike (S/2003 J6)	1.3 (2)	0.156	2003
29. Hermippe	4.8 (7.8)	0.210	2001
30. Thelxinoe (S/2003 J22)	0.6 (1)	0.221	2003
31. Ananke	8.8 (14)	0.244	1951
Carme retrograde group			
32. S/2003 J15	0.6 (1)	0.110	2003
33. Eurydome	2 (3)	0.276	2001
34. S/2003 J17	0.6 (1)	0.190	2003
35. Pasithee	0.6 (1)	0.267	2001
36. S/2003 J10	0.6 (1)	0.214	2003
37. Chaldene	1.3 (2)	0.251	2000
38. Isonoe	1.3 (2)	0.246	2000

Moon	Radius (miles [km])	Orbital Eccentricity	Year Discovered
39. Erinome	1 (1.5)	0.266	2000
40. Kale	0.6 (1)	0.260	2001
41. Aitne	1 (1.5)	0.264	2001
42. Taygete	1.5 (2.5)	0.252	2000
43. Kallichore (S/2003 J11)	0.6 (1)	0.264	2003
44. Eukelade (S/2003 J1)	1.3 (2)	0.272	2003
45. Arche (S/2002 J1)	1 (1.5)	0.259	2002
46. S/2003 J9	0.3 (0.5)	0.269	2003
47. Carme	15 (23)	0.253	1938
48. Kalyke	1.5 (2.5)	0.245	2000
Pasiphae irregular group			
49. Sponde	0.6 (1)	0.312	2001
50. Magaclite	2 (3)	0.421	2000
51. S/2003 J5	1.3 (2)	0.210	2003
52. S/2003 J19	0.6 (1)	0.334	2003
53. S/2003 J23	0.6 (1)	0.309	2003
54. Hegemone (S/2003 J8)	1 (1.5)	0.328	2003
55. Pasiphae	18 (29)	0.409	1908
56. Cyllene (S/2003 J13)	0.6 (1)	0.319	2003
57. S/2003 J4	0.6 (1)	0.204	2003
58. Sinope	12 (19)	0.250	1914
59. Aoede (S/2003 J7)	1.3 (2)	0.432	2003
60. Autonoe	1.3 (2)	0.334	2001
61. Calirrhoe	2 (3.5)	0.283	1999
62. Kore (S/2003 J14)	0.6 (1)	0.325	2003
63. S/2003 J2	0.6 (1)	0.380	2003

(continued from page 74)

the distance from Boston to New York City, with mountains more than twice the height of Mount Everest (which is about 5.5 miles high)! Small bodies have small gravity, and so higher mountains can exist without being pulled down by gravity. Still, there are no good working hypotheses for what could cause such large mountains to form on such a small body.

Amalthea's surface is dark and reddish. The surface may be that color from its own composition combined with a constant bombardment of energetic particles funneled through Jupiter's magnetic field, or it may be coated with sulfur from neighboring Io's incredibly active volcanoes. Even more strange, steep slopes on Amalthea appear bright green. Green is an uncommon color for solar system bodies, and why it appears on Amalthea is unknown.

4. Thebe

Thebe is named for a water nymph and is another small, irregularly shaped moon with little known about it. Stephen Synnott discovered Thebe in 1979 from images takes by *Voyager*.

Images of Jupiter's four small inner satellites were taken by Galileo. From left to right, in order of decreasing distance to Jupiter: Thebe, Amalthea (the largest moon), Adrastea (the smallest), and Metis, presented on the same scale. Adrastea and Metis were first resolved by a spacecraft camera in these images. The bottom panel shows computer models of the moons' shapes. (NASA/ JPL)

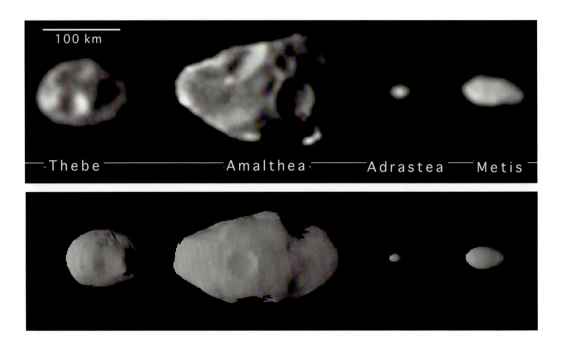

100 km

Thebe — Amalthea — Adrastea — Metis

In the images, Thebe appears as a tiny black dot against the bright, swirling background of Jupiter's clouds.

5. Io

In ancient Greek religion, Io was a princess beloved of Zeus, who turned her into a white heifer to disguise her from Hera, his goddess wife. Quite unlike the passive heifer, Io the moon is one of the most interesting and exciting objects in the solar system. Io, seen in the photo below, is the most volcanically active body in the solar system, and aside from Earth, the only other body on which man has actually witnessed volcanic activity. There are 150 to 300 volcanic hot spots on Io, along with lava lakes and caldera-like depressions called paterae (patera is a planetary geology term for a shallow, irregular crater). See the sidebar "Fossa, Sulci, and Other Terms for Planetary Landforms" on page 170. Io has huge volcanic plumes and visible lava lakes, as well as mountains 52,000 feet (16 km) high, twice as high as Mount Everest. On Io, the mountains may actually be formed in ways similar to the ways mountains are formed on Earth: There is photographic evidence for *plate tectonics* of a sort, with *subduction* and thrusting of crustal plates forming mountains. Io's mountain Mongibello Mons is higher than any mountain in North America.

The Galileo spacecraft acquired its highest resolution images of Io on July 3, 1999. (NASA/JPL)

Io has valleys as much as 2.5 miles (4 km) deep, 200 dark fresh lava flows, and 100 active volcanic hot spots. Io is so volcanically active that it is constantly sending lighter elements into its atmosphere, where they are stripped away by Jupiter's magnetic field. This process may be what has made Io the densest body in the outer solar system. Though Io is just a little bit larger than Earth's Moon, it has a metallic core that may make up 50 percent of its radius, and its heat output is even greater than the Earth's. Some of Io's surprising physical char-

FUNDAMENTAL FACTS ABOUT IO

polar radius	1,133 miles (1,815 km)
equatorial radius	1,137 miles (1,830 km)
albedo	0.63
density	222 pounds per cubic feet (3,528 kg/m^3)
gravity	5.94 feet per second squared (1.81 m/sec^2)
orbital period around Jupiter	1.769 Earth days
magnetic field	10^{-6} Tesla

acteristics are listed in the table here, including its high density and commensurately high gravity and its magnetic field strength.

Galileo Galilei discovered Io in 1610, in the same week that he discovered Ganymede, Europa, and Callisto. Io orbits Jupiter four times for every time Ganymede orbits Jupiter once, as well as four times for every two times that Europa orbits the planet. This arrangement made it possible to estimate the moons' masses long before spacecraft visited them. Io orbits Jupiter at just 5.9 Jupiter radii. From Io, Jupiter would look 40 times larger than the Moon does from Earth. The *Voyager 1* mission first discovered Io's volcanism in March 1979. *Galileo,* the next mission slated to visit the outer solar system, was already planned. The discovery of active volcanoes was so startling that *Galileo* was promptly redesigned to include an instrument for infrared mapping, so that the surface temperatures of the moon could be detected. *Galileo* orbited Jupiter 34 times and took 600 images of Io, some at a resolution of around 33 feet (10 m) per pixel. Gas and dust plumes 500 miles (800 km) high issue from the volcanoes. These are truly gigantic plumes; the plume from a recent large eruption in the Kamchatka peninsula on Earth was only three miles (5 km) high. (The Kamchatka peninsula reaches south from eastern

Siberia.) One plume that erupted from Io, shown in the figure below, created fallout that covers an area the size of Alaska. The vent is a dark spot just north of the triangular-shaped plateau (right center). To the left, the surface is covered by colorful lava flows rich in sulfur.

In one image two volcanic plumes were captured on Io (see image page 82). One plume was captured on the bright limb or edge of the moon, erupting over the caldera named Pillan patera. Pillan patera's plume is 85 miles (140 km) high and has also been seen by the *Hubble Space Telescope*. The second plume, Prometheus, is seen near the terminator (the boundary between day and night). The shadow of the airborne plume can be seen extending to the right of the eruption vent. (The vent is near the center of the bright and dark rings). The Prometheus plume can be seen in every *Galileo* image viewing the correct part of Io, as well as by every *Voyager* image of that region. It is possible therefore that this plume has been continuously active for more than 18 years.

The lavas produced by the voluminous and possibly continuous eruptions on Io create rivers of flowing lava on the moon's surface, such as those shown in the photo on page 83. The lava channel is dark and runs to the right from the

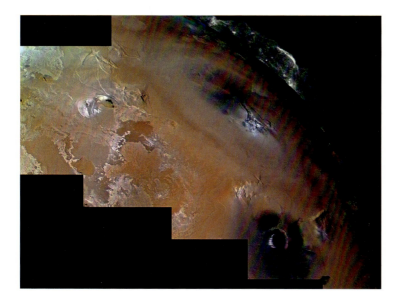

The plume from Io's volcano Pele rises 190 miles (300 km) above the surface in an umbrella-like shape and its fallout covers an area the size of Alaska. (NASA/JPL)

dark patera (Emakong patera) at the left of this mosaic. As on Earth, lava cools as it flows along the surface, and so maintaining surface lava channels hundreds of kilometers long is a problem. Close inspection of this and other lava channels on Io show that they are partially roofed over: The cooling lava has formed a *crust* that connects to the sides of the channel, forming a roof under which lava can flow and retain its heat through insulation.

Io's surface is sulfur-covered from volcanic eruptions that spray different compositions onto its surface, creating bright regions colored white, yellow, orange, red, green, and black, depending on the sulfur content and chemical composition. The different sulfur compounds required to make the many colors of Io's surface can be created by differing amounts of heat. Shown in the photo on page 83, the small SO_2 plumes make irregular or circular rings 30 to 220 miles (50 to 350 km) in diameter that are white, yellow, or black. Giant sulfur plumes make oval markings on Io's surface, elongated north to south, 220 to 500 miles (350 to 800 km) in length, colored red or orange. These are huge extents of eruption: On the Earth, an eruption 500 miles (800 km) in length would reach from Boston to Washington, D.C. Some of these giant plumes erupt fairly continuously, while others seem to be single events.

Because Io orbits within the most intense part of Jupiter's magnetosphere, Jupiter's magnetic field is strong enough to pull atmospheric material and volcanic gases away from Io and into orbit around Jupiter (Io, surprisingly, has its own magnetic field, about 10^{-6} Tesla, but it is much weaker than Jupiter's).

Both Pillan and Prometheus paterae are erupting in this image, as described in the text. (NASA/JPL)

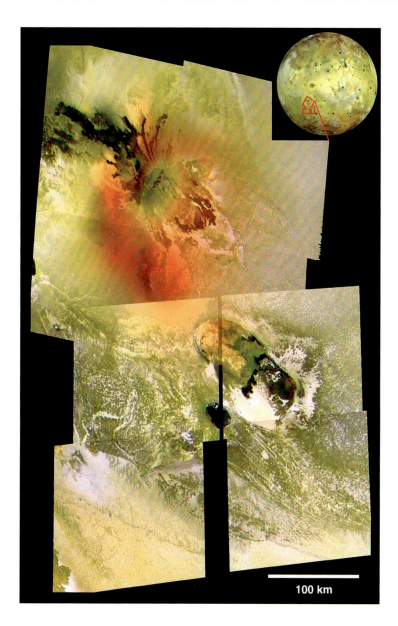

Culann Patera (top center) is one of the most colorful volcanoes on Jupiter's innermost large moon, Io. The volcano has produced dark black and red lava flows as well as diffuse rings of red and yellow sulfur particles. (NASA/JPL/Galileo)

100 km

About one ton (1,000 kg) of sulfur, atmospheric sodium, potassium, and chlorine is pulled off Io every second, and these ions form a cloud around the planet as well as a doughnut-shaped cloud of ions around Jupiter, in Io's orbit. These ions create spectacular auroras by interacting with Jupiter's magnetic field. The source of the sodium, potassium, and chlorine

is not understood, however, because these volatile elements should have been depleted earlier in Io's history. The movement of these electrically charged ions through Jupiter's magnetic field creates the effect of a electrical generator: Every magnetic field can create a linked electrical field, and, in this case, Io and its charged particles become the electrical field. Io commonly develops 400,000 volts across its diameter and generates an electric current of 3 million amperes that flows along the magnetic field to Jupiter's ionosphere.

At first volcanism was thought to be entirely sulfuric, but temperatures now show differently. Sulfur eruptions occur at temperatures less than about 800°F (430°C), while basaltic lava eruptions on Earth occur at temperatures of about 2,000 to 2,700°F (1,100 to 1,500°C). The *Voyager 1* mission measured eruption temperatures only up to about 750°F (400°C), consistent with sulfur-rich eruptions, but the Galileo mission detected eruptions at temperatures greater than 800°F (430°C), and even saw one eruption in 1997 at Pillan patera with lava temperatures of 2,900 ± 45°F (1,600 ± 25°C). Four eruptive centers have been measured at above 2,060°F (1,100°C): Pillan, Masubi, Pele, and Surt paterae. These high eruptive temperatures show that silicate liquids must be involved in some of the eruptions, while others with lower temperatures consist mainly of sulfur compounds. Most volcanic eruptions on Io are at about 2,600°F (1,430°C), about the temperature of hot silicate magmas on Earth. The hottest eruptions on Io, at about 2,900°F (1,600°C), are hotter than any volcanic eruptions on Earth now, and may be hotter than any eruptions that ever occurred on Earth. This could be due to two things: Either lava is heated well above the temperature it requires to become molten (superheated), or the material being melted requires very high temperatures to melt, higher than the *mantle* on Earth requires. The images in the photo on page 86 from the Galileo mission show changes in the largest active lava flows in the solar system, in the 190-mile (300-km)-long Amirani lava field on Io. Researchers have identified 23 distinct new flows by comparing the two images taken 134 days apart, on October 11, 1999, and February 22, 2000. Individual flows within it are each several kilometers or miles long, which is about the size of the entire active eruption on Kilauea, Hawaii. The new lava flows at Amirani identified in these images cover about 240 square miles (620 km^2) in total over the five months of observation. In comparison, Kilauea covered only about four square miles (10 km^2) in the same time, and Io's Prometheus lava flow field covered about 24 square miles (60 km^2).

The largest eruptive center on Io is called Loki patera. Loki patera is 140 miles (220 km) long, with an area of 11,600 square miles (30,000 km^2). Loki is covered with dark silicate lava from recent eruption, and at its eruptive height, it can produce more than 10 percent of the moon's total heat output. There are enough eruptions from Loki and other patera that the surface age of Io is current (meaning that the whole surface is freshly made); in fact, there were eight volcanic eruptions during *Voyager 1* flyby alone. The record of cratering seen on other solar system bodies cannot be seen on Io, as they have all been covered by the volcanic eruptions. In five years of observations, *Galileo* saw more than 16 SO$_2$ plumes making

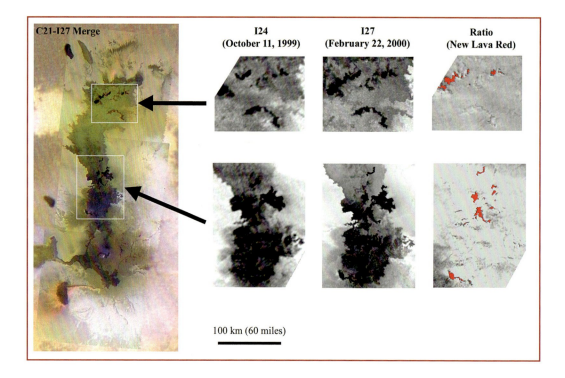

C21-I27 Merge

I24 (October 11, 1999)

I27 (February 22, 2000)

Ratio (New Lava Red)

100 km (60 miles)

Over five months of observation, Io's Amirani lava field produced about 240 square miles (620 km²) of new lava; in comparison, Earth's Kilauea volcano in Hawaii produced only four square miles (10 km²). (NASA/ JPL/Galileo)

rings on Io's surface and six giant plumes consisting mainly of sulfur alone. During these five years, 17 percent of Io's surface was resurfaced by volcanic activity. More than 100 active volcanic centers have been found. Compared to the Earth, where every year around eight to 15 km³ of lava is erupted onto the surface in total from all volcanoes, over 500 km³ of lava is erupted onto the surface of Io every year. Io's mass is tiny, only about 1.5 percent of the Earth's mass, but it erupts over 30 times as much lava. Io is so voluminously volcanic that if it had erupted at this rate continuously over the last 4 billion years, it would have melted its mantle and crust 80 times over and erupted 40 times its own volume! This indicates a great ability of the silicate portion of Io to flow, constantly reforming a spherical planet.

This kind of volcanic recycling is unheard of in planetary science, and it requires both very high rates of heating and an unusual planetary composition. On the Earth, the main constituent of the upper mantle is the mineral *olivine*, and olivine is also a large component in erupted lavas. Analysis of reflectance spectra from Io indicate that the lavas are very high in magnesium; this helps explain their high temperatures, because magnesium-rich minerals have high melting temperatures in general. Surprisingly, though, the spectra indicate that the main mineral is orthopyroxene, and not olivine. This result is reminiscent of Earth's Moon, where a high percentage of its mantle is also thought to be made of orthopyroxene. One possible explanation for Io's strange composition is that some elements may have been vaporized away from hot magmas throughout Io's history, and therefore depleted from the planet as a whole (or trapped in cold areas near Io's poles). Elements likely to be vaporized are also those that lower the melting temperature of the mantle, such as sodium, potassium, and iron, and if these have been lost throughout geologic history from the planet, the temperatures required to melt the planet's interior would have risen through time.

Each day the surface temperature of Io varies between -330 and -200°F (-200 and -140°C). *Voyager 1* measured 80°F (27°C) in a dark region, exceptionally warm compared to the rest of

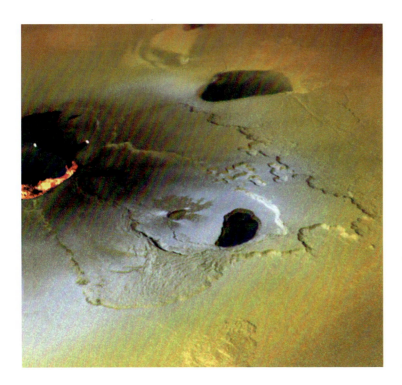

Galileo captured a dynamic eruption at Tvashtar Catena, a chain of volcanic bowls on Jupiter's moon Io. (NASA/JPL/University of Arizona)

the planet. Io's surface temperatures, with the exception of volcanic hot spots, are about the same on the poles as at the equator. On Earth, the equatorial areas are hotter because of the more direct sunshine. Why is Io different? There may be more volcanic heating at the poles, or there may be materials there that hold heat more effectively, keeping the poles warm while the equator cools. To support the volcanoes that are seen, the crust has to be a few tens of kilometers thick. Io thus has another paradox: It must have a thick stiff crust to support the high volcanoes that exist, but all of its material needs to flow enough so that the moon has constant volcanic eruptions and yet remains a sphere.

Tidal stresses from Jupiter's giant, close-by mass are thought to be important in production of volcanism. While Io is held tightly in Jupiter's gravity field, the other moons that pass by Io pull it in other directions. Because of these strong and opposing pulls, the surface of Io oscillates by hundreds of meters as it rotates. This kind of wild oscillation of a solid can create significant amounts of heat through friction, and it has been thought to create the volcanic activity seen. Tidal stresses are not the final word on the source of heat for Io's eruptions, however, now that more detailed calculations of heating have been made. Io's power output from volcanism is about 2.5 W/m^2. This value may be more than the energy contributed from Jupiter through tidal heating. (Some researchers contend that it is more than twice the energy that Jupiter contributes.) It is about twice the magnitude of the heating provided by electromagnetic heating from the ion storm around Io, and it also exceeds any heat possible from radioactive decay. Io's heat output, in fact, is more than twice the Earth's. Tidal heating is still accepted as the method for creating the heat required for the extravagant quantity of volcanic activity on Io.

Some insight into the internal structure of a planet or moon can be obtained from its moment of inertia factor. The moment of inertia factor is a measure of how much force is required to speed up or slow down the body's spin. (For more on moment of inertia, see the sidebar "Moment of Inertia" on page 94.) Io's moment of inertia factor is 0.37685 ± 0.00035. Based on its moment of inertia, it is thought to be rocky and silicate-rich, and its average density is estimated at 1,228 lb/yd³ (3,527.8 kg/m³), making it the densest object in the outer solar system. By comparison, Europa's average density is 190 lb/ft³ (3,000 kg/m³), and ice-rich moons Ganymede and Callisto have densities of 121 lb/ft³ (1,900 kg/m³) and 114 lb/ft³ (1,800 kg/m³). The mantle of Io is probably dominated by olivine, just as the Earth's mantle is, based on density and temperature constraints. Of the Galilean satellites, only Io and Europa are thought to have olivine-rich mantles. The density of an H chondrite, thought to be one of the most primitive materials in the solar system, is 243 lb/ft³ (3,850 kg/m³), and so Io and the other outer planets have been depleted of their heavy elements.

Io's moment of inertia factor also requires that it have a core. The core of Io is thought to be a combination of iron (Fe) and sulfur (S) in two forms, either Fe and FeS, or Fe_3O_4 and FeS. The iron compounds here differ only by how much oxygen (O) they have bonded with; the oxygen state of the inside of Io is not known. Sulfur makes the core material less dense, and thus allows a larger core while still matching the moment of inertia factor for the planet. Depending on the core's composition, and hence its density, Io's core may be from 38 to 53 percent of the satellite's radius. The core is thought to be overlain by a silicate mantle, and then about 50 miles (80 km) of crust, with a density of about 190 lb/ft³ (3,000 kg/m³).

Perhaps Io is still a mushy magma ocean, covered with a crust. This structure would complicate the explanation for the moon's weak magnetic field: If Io does consist primarily of a magma ocean, it might lack the internal structure thought necessary to form a magnetic field. On the other hand, a magma ocean might remain well mixed through convection, helping to explain the homogeneity of the lava compositions erupted on Io. Density calculations indicate that the mantle cannot be fully molten, though it may be partially molten. Io has become a focus of tremendous scientific attention, and it is hoped that spacecraft may send back data to help answer some of these questions.

6. Europa

Though Io and Europa are neighbors orbiting Jupiter, the two moons are extremely different. Io has volcanoes and mountains, and Europa has none. Europa has a significant ice fraction in its composition, while Io has none. Europa's surface is younger than Callisto's and Ganymede's, and its icy surface makes it one of the brightest objects in the solar system: Its *albedo* is 0.64; in comparison, the Moon's albedo is only 0.07, though Neptune's moon Triton has an albedo of 0.7. Like Io, Europa also has a small magnetic field, about 10^{-7} Tesla.

Europa is named after a lovely Phoenician princess from Greek mythology. Zeus saw Europa gathering flowers and immediately fell in love with her. Zeus transformed himself into a white bull and carried Europa away to the island of Crete. He then revealed his true identity and made Europa the first queen of Crete. Her sons fought in the Trojan War against Greece. Zeus later recreated the shape of the white bull in the stars, which is now known as the constellation Taurus.

Europa's surface is relatively flat, having no very high mountains or great basins, but it does display linear surface features, ridges, and troughs, both singly and in sets ("en echelon" in geologic nomenclature), along with craters and chaos terrain. Chaos terrains are circular or elliptical regions where the lineations of the icy crust have been erased and replaced with jumbled, broken ice. In places, early flat plains are still visible, but 22 percent of the surface has been resurfaced with chaos terrain, and 30 percent by other tectonic activity. (The early plains can be identified simply by the fact that the other tectonic features are clearly placed on top of and in these plains.) Fifty percent of Europa has been resurfaced in the last 50 to 100 Ma, which is a very short time geologically speaking. Like Io, therefore, Europa has a surface currently being resurfaced, though with ice in its case rather than molten rock.

Two examples of chaos terrain are shown in this book. In the first, shown on page 93, irregular blocks of water ice were formed by the shattering and movement of the existing crust. The blocks were shifted, rotated, tipped, and partially submerged within a mobile material that was either liquid water, warm mobile ice, or an ice and water slush. The presence of young fractures cutting through this region indicates that the surface froze again into solid, brittle ice. This first example of chaos terrain has blocks of ice with straight lines and the angles associated with brittle fracture, but this second feature shows round shapes and curves. In the second example, Thera and Thrace are two dark, reddish regions of enigmatic terrain that disrupt the older icy ridged plains on Europa as seen in the photo on page 92). Thera (left) is about 43 × 53 miles (70 × 85 km) and appears to lie slightly below the level of the surrounding plains. In contrast, Thrace (right) is longer, shows a hummocky texture, and appears to stand at or slightly above the older surrounding bright plains.

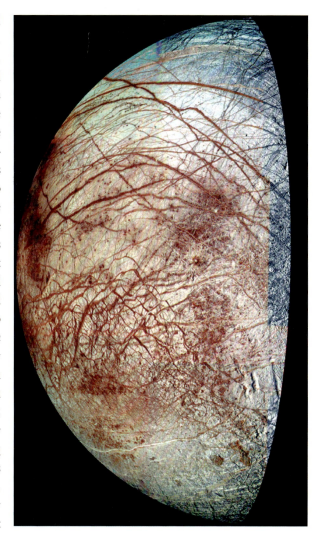

The dark brown areas on Jupiter's moon Europa are material from the interior or implanted by impact; the dark blue areas are coarse-grained ice; and the paler blue areas are fresh, fine-grained ice. (NASA/JPL/ University of Arizona)

Europa is thought to have an iron core, a silicate mantle, a layer of liquid water, and a crust of ice. The liquid ocean under the crust of ice could explain the lineations in the crust: Movement of the fluid beneath created tension and compression in the crust, making it buckle or fault. Europa's surface is covered with long curved features called arcuate ridges, unique in the solar system. The longest arcuate ridge is called Astypalaea Linea, and it is over 560 miles (900 km) long. Wide, overlapping cracklike features also may indicate a kind of ice plate tectonics, in which certain areas spread and new

This chaos region on Europa is thought to have been made by movement in a partly solid, partly liquid, water crust. (NASA/JPL/Galileo)

ice forms in the cracks, while in other areas the icy plates press together and make long compressional ridges.

These long cracks may be caused by gravitational pulls from Jupiter and from the other Galilean satellites. Intriguingly, this particular process would only work if the icy shell were floating on a liquid ocean. Since water flows so much more easily than solid ice, it can respond to tidal forces more quickly and it can move farther than solid materials would. If Europa has a subsurface sea, then its ice crust should rise and fall by 100 feet (30 m) during each 3.6-day orbit around Jupiter. These wild fluctuations are sufficient to cause the cracking and ice resurfacing seen on the planet. If there is no liquid ocean, tides will only cause the surface to rise and fall by three feet (1 m), insufficient to cause the lineations.

Different approaches to calculating the thickness of the icy shells and oceans on Europa bring different predictions. Using gravity measurements from the Galileo mission, a team of scientists lead by Frank Sohl at the University of Münster concluded that Europa has an ice crust between 75 and 105 miles (120 and 170 km) thick. Its core is constrained to be between 10 and 45 percent of the its radius, and its mantle is thought to be silicate and dominated by olivine, as Io's mantle is. A separate team also at the University of Münster, led by Hauke Hussmann, created computer models of the tidal heating created in Europa by Jupiter's giant gravity field and determined that Europa's ice shell can be only a few tens of kilometers thick, with underlying water oceans about 60 miles (100 km) deep. Hussmann's estimate, then, gives Europa an ice shell only about half, or less, than the one Sohl's estimate gives. Either of these estimates indicates that Europa has more water than the total on the planet Earth!

Some scientists suggest that heat from below breaks up blocks of ice at the surface, which drift around and agglomerate to form Europa's surface features. The heat from below comes from radioactivity and tidal heating, while the surface is cooling everywhere. The oceans are thought to be convecting fiercely enough to have no stratification, in contrast to Earth's oceans, which are stratified. Other scientists believe that Europa's oceans are maintained as liquid and not allowed to freeze by tidal friction heating from Jupiter.

The round shapes in Europa's Thera and Thrace regions may have been caused by warmer liquid rising from below and melting through the solid crust. (NASA/JPL/Galileo)

Chaos regions, such as the Thera region and the Thrace region, may also be caused by dynamics of a liquid ocean under an ice crust. Heating in the solid interior of the planet from tides and radioactive elements may cause icy blobs buoyant enough from warming to rise through the liquid ocean.

MOMENT OF INERTIA

The moment of inertia of a planet is a measure of how much force is required to increase the spin of the planet. (In more technical terms, the angular acceleration of an object is proportional to the torque acting on the object, and the proportional constant is called the moment of inertia of the object.)

The moment of inertia depends on the mass of the planet and on how this mass is distributed around the planet's center. The farther the bulk of the mass is from the center of the planet, the greater the moment of inertia. In other words, if all the mass is at the outside, it takes more force to spin the planet than if all the mass is at the center. This is similar to an example of two wheels with the same mass: one is a solid plate and the other is a bicycle wheel, with almost all the mass at the rim. The bicycle wheel has the greater moment of inertia and takes more force to create the same angular acceleration. The units of the moment of inertia are units of mass times distance squared; for example, lb × ft^2 or kg × m^2.

By definition, the moment of inertia I is defined as the sum of mr^2 for every piece of mass m of the object, where r is the radius for that mass m. In a planet, the density changes with radius, and so the moment of inertia needs to be calculated with an integral:

$$I = \int_0^{r_f} \left(\rho(r) \right) r^2 dr,$$

where r_o is the center of the planet and r_f is the total radius of the planet, $\rho(r)$ is the change of density with radius in the planet, and r is the radius of the planet and the variable of integration. To compare moments of inertia among planets, scientists calculate what is called the moment of inertia factor. By dividing the moment of inertia by the total mass of the planet M and the total radius squared R^2, the result is the part of the moment of inertia that is due entirely to radial changes in density in the planet.

These blobs, called diapers, would cause disruption of the crust when they strike it from below, creating chaos regions. Smaller diapers may cause other small circular features, called lenticulae. Lenticulae are named after the Latin word for freckles, and may be domes, pits, or dark spots. In the image

This division also produces a non-dimensional number because all the units cancel. The equation for the moment of inertia factor, K, is as follows:

$$K = \frac{I}{MR^2}.$$

The issue with calculating the moment of inertia factor for a planet is that, aside from the Earth, there is no really specific information on the density gradients inside the planet. There is another equation, the rotation equation, that allows the calculation of moment of inertia factor by using parameters that can be measured. This equation gives a relationship between T, the rotation period of the planet; K, the moment of inertia factor of the planet; M, the mass of the planet; G, the gravitational constant; R, the planet's polar radius; D, the density for the large body; a, the planet's semimajor axis; i, the orbital inclination of the planet; m, the total mass of all satellites that orbit the large body; d, the mean density for the total satellites that orbit large body; and r, the mean polar radius of all satellites that orbit the large body:

$$T^2 = K\left(\frac{4\pi^2 D^3}{G(M+m)\cos^2 i}\right)\left(\left(\frac{m}{M}\right)\left(\frac{r}{R}\right) + \left(\frac{M}{n}\right)\left(\frac{D}{d}\right)\right)(4\pi a r K).$$

By getting more and more accurate measures of the moment of inertia factor of Mars, for example, from these external measurements, scientists can test their models for the interior of Mars. By integrating their modeled density structures, they can see whether the model creates a moment of inertia factor close to what is actually measured for Mars. On the Earth, the moment of inertia factor can be used to test for core densities, helping constrain the percentage of light elements that have to be mixed into the iron and nickel composition.

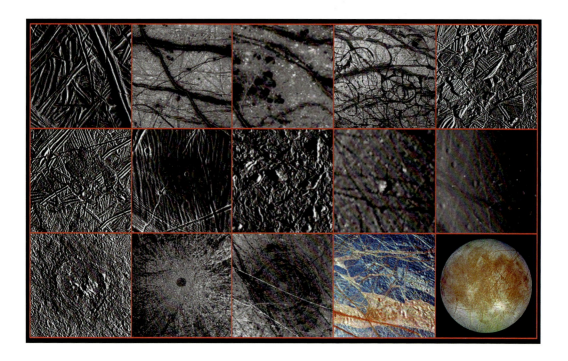

These 15 images show the range of surface images on Europa. (NASA/JPL/Galileo)

presented above, a wide variety of surface feature examples from Europa are shown, including chaos regions, lineations, lenticulae, and craters modified by flowing ice.

The discovery of a liquid water ocean on Europa is immensely exciting, since the combination of liquid water and organic molecules are what made life possible on Earth, so perhaps Europa, with those same elements, may also produce life. As Richard Terrile, a NASA scientist, has pointed out, "How often is an ocean discovered? The last one was the Pacific, by Balboa, and that was five hundred years ago!" Two experts in cratering, Elisabetta Pierazzo at the University of Arizona and Christopher Chyba of Stanford University, have calculated that at low impact velocities (10 mi/sec, or 16 km/sec), the carbon and carbon-based molecules held in comets are retained on Europa and not expelled back into space. Over solar system history, large comets may have delivered 1 to 10 billions of tons of carbon to Europa. This is a few times more carbon than is contained in the upper 60 feet (200 m) of Earth's oceans, where most life exists, but it is about 100 times less than all the carbon in all the oceans when the entire ocean depth is considered. Another researcher, Paul M. Schenk at the Lunar and Planetary Institute, measured the depths of impact craters on Europa, Ganymede, and Callisto, and found that Europa's crater shapes indicate an especially dense and cold ice shell, indicating that exchange of organic materials between the surface and the liquid ocean beneath could be slow or even impossible.

Because Europa is a possible place to find life, planetary scientists are especially careful not to allow any spacecraft from Earth to contaminate the moon (no matter how clean the spacecraft is when it leaves Earth, and no matter how long its passage in space, a spacecraft is almost certain to carry microscopic life with it from Earth). Europa is the main reason that NASA crashed the Galileo mission into Jupiter; the scientists wanted to leave no chance that the spacecraft might fall into Europa as its communications and steering failed at the end of its lifetime.

7. Ganymede

In ancient Greek mythology, Ganymede was a young sheepherder beloved of Zeus. Ganymede is the largest planetary moon in the solar system, with a radius of 1,646 miles (2,634 km), making it larger than both Pluto and Mercury. Despite its huge size, its core is thought to be unusually small in comparison with its radius, at only 25 to 30 percent of the radius. The remainder of Ganymede's radius is taken up by a very thick mantle, thought to be half silicate minerals and half ice, covered with a thin icy crust.

Ganymede has its own magnetic field, about 2×10^{-6} Tesla, though it is thought to be largely icy and to contain less rock than others of Jupiter's moons because of its low density. The magnetic field on Ganymede is a mystery. Though models indicate that Ganymede's interior should have been molten at one time, and therefore capable of convecting rapidly and creating a dynamo-driven magnetic field, Ganymede should have long since cooled and solidified completely. Tidal flexing from resonant orbits with Europa and Io may contribute to heating and convection, but no clear model for the production of a magnetic field exists.

Ganymede's mantle is thought to be between 550 to 700 miles (900 to 1,100 km) thick and probably made of ices, although it is possible to make a model of Ganymede with a silicate mantle that is still consistent with the density and moment of inertia measurements. This mantle shell is probably subdivided into thinner shells created by different phases of ice. Water ice metamorphoses into different crystal structures depending on pressure: Humans are most familiar with ice I, but with greater pressure, at the temperatures of Ganymede, ice transforms into ice III, ice V, and ice VI. These are called polymorphs: the same materials but with different crystal structures, as shown in the figure on page 99.

At low pressures the water molecules organize themselves according to charge. The two hydrogen atoms in a water molecule are slightly positively charged, and the oxygen is slightly negatively charged. A hydrogen from a neighboring water molecule will therefore weakly bond with the oxygen, and the hydrogens will themselves weakly bond with oxygens in other molecules. Because water molecules are shaped like boomerangs, with the oxygen at the bend in the boomerang and a hydrogen at each end of the boomerang making an angle of about 108 degrees, the water molecules make a honeycomb shape when they all weakly bond together to make ice I.

The honeycomb structure of ice I is weak, not only because the water molecules are not efficiently packed together, but also because the electrical bonds between molecules are weak. As pressure is increased on ice I beyond the strength of the weak intermolecular bonds, the molecules

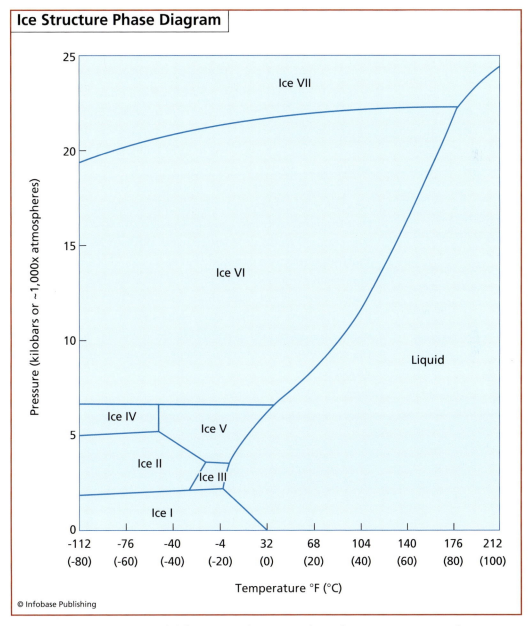

Water freezes into a variety of different crystal structures depending on its pressure and temperature conditions.

eventually are forced into more and more efficient packing schemes. Pressure also inhibits melting, and so the higher-pressure polymorphs of ice can exist at temperatures above 32°F (0°C).

Ganymede's surface records many processes. The surface has both dark and bright areas; the interface of a bright and dark area is shown in the figure below. The dark areas are thought to be more ancient, in part because they are heavily cratered. The dark areas may also have more carbonaceous dust covering them, perhaps from meteorite impacts.

The bright parts of Ganymede's surface are dominated by sets of straight or slightly curving troughs. The term *sulcus*, meaning "a groove or burrow," is often used to name these features. The upper image on page 101 shows a close-up of an area of the Harpagia Sulcus region. The grooves can be thousands of miles in length and hundreds of yards deep. The team of Andrew Dombard and William McKinnon from Washington University find that extensional stresses in an ice crust cause a weak lower layer to flow, while a cold, strong upper layer breaks open along parallel lines. The model predicts that the icy shell of Ganymede split open in tension, and mushy ice flowed up to partly fill the tensional cracks. This is the currently most plausible theory for the formation of these grooves.

Ganymede's bright and dark areas meet in this image from the Galileo mission. The ancient, dark terrain of Nicholson Regio (left) shows many large impact craters, and zones of fractures oriented generally parallel to the boundary between the dark and bright regions. In contrast, the bright terrain of Harpagia Sulcus (right) is less cratered and relatively smooth. (NASA/JPL/Galileo)

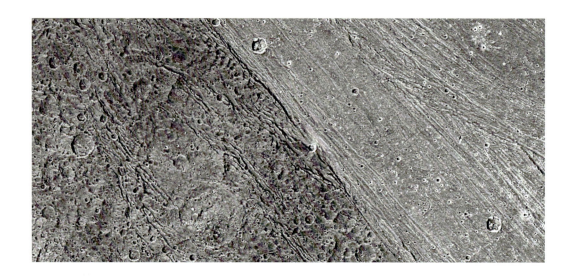

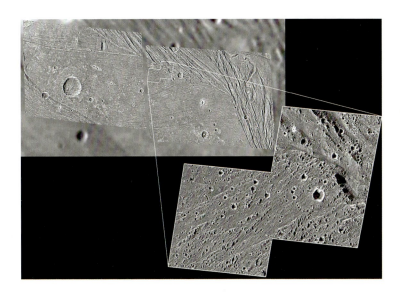

This high-resolution image of Ganymede's Harpagia Sulcus shows features as small as 52 feet (16 m) across. (NASA/JPL/Galileo)

The large craters on Ganymede exist as outlines rather than depressions; they seem to have been filled in. The ejecta from craters on Ganymede's dark regions is generally dark itself, and from craters on the light regions the ejecta is bright. These flat craters are consistent with cratering on an icy surface, which over many hundreds of millions of years will slowly rise and readjust to make a nearly flat surface from a cratered depression. (See the sidebar "Rheology, or How Solids Can Flow" on page 106.) These ancient erased craters are called palimpsests, a term referring to antique, reused papers and papyruses on which older writing was still visible underneath newer writing. Palimpsests on Ganymede range from 30 to 250 miles (50 to 400 km) in diameter. In the lower *Galileo* image on the opposite page, the central depressed feature is a caldera, a collapsed volcanic feature. The caldera was never round, as almost all impact craters are. Surrounding the caldera are several impact craters in varying degrees of their transitions to palimpsests.

Ganymede and Europa share some similar surface banding. The images shown on page 103, taken by *Galileo,* show a same-scale comparison between Arbela Sulcus on Ganymede (left) and an unnamed band on Europa (right). Arbela Sulcus is one of the smoothest lanes of bright terrain identified on Ganymede and shows very subtle striations along its length.

Arbela contrasts markedly from the surrounding heavily cratered dark terrain. On Europa the scarcity of craters illustrates the relative youth of its surface compared to Ganymede's. Arbela Sulcus is an unusual feature for Ganymede and may have formed by complete separation of Ganymede's icy crust, like bands on Europa. Prominent fractures on either side of Arbela appear to have been offset by about 40 miles (65 km) along the length of the band, suggesting that strike-slip faulting occurred during the formation of Arbela Sulcus.

In this second image of a portion of Arbela Sulcus (see figure on page 104), its bright terrain is the youngest surface, slicing north-south across the image. To the right is the oldest terrain in this area, rolling and relatively densely cratered Nicholson Regio; on the left is a region of highly deformed grooved terrain, intermediate in relative age.

Ganymede seems to be completely inactive now, with no new grooves or ice flows. Some scientists estimate that Gany-

The shallow, scalloped depression in the center of this image is a caldera on Ganymede, surrounded with impact craters, and likely marks a place where liquid once lay beneath the surface. (NASA/JPL/ Galileo)

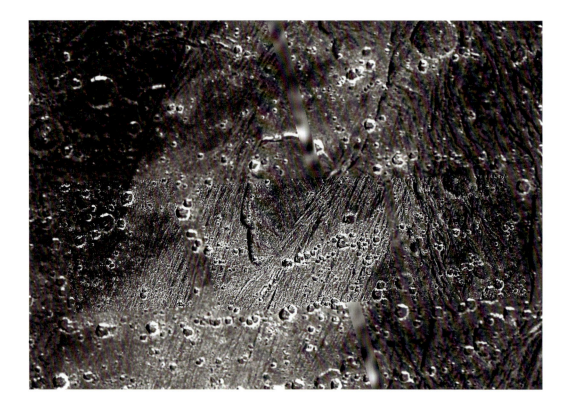

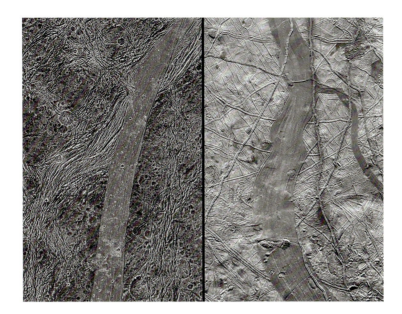

Europa's young surface (right) contrasts with Ganymede's more cratered surface. (NASA/JPL/Galileo)

mede was only active during the first billion years of the solar system, but future space missions will be looking closely at the surface for recent activity.

8. Callisto

With a radius of about 1,500 miles (2,400 km), Callisto is about the size of Mercury. It is the third largest moon in the solar system, after Jupiter's Ganymede and Saturn's Titan. Callisto orbits just outside Jupiter's main radiation belt from its magnetic field and is therefore relieved of the magnetic storms and ionic stripping that Io experiences. Callisto, like the other Galilean satellites, does produce its own magnetic field of about 4×10^{-9} Tesla, probably from circulating salty water.

Callisto is named after a woman from Greek myth who was either the daughter of Lycaon, king of Arcadia, or a river nymph. Like the women other moons of Jupiter are named after, Callisto was pursued by Zeus. Zeus fell in love with her astonishing beauty and disguised himself as Apollo to win her over. Hera, Zeus's wife, took revenge on Callisto by turning her into a bear, and in the end, Zeus placed her and her son in the heavens as the constellations Ursa Major and Ursa Minor.

Ganymede's surface is exceptionally complex both in features and age. The three regions in this image move from youngest on the left to oldest on the right. (NASA/JPL/Galileo)

Callisto has an icy, cratered, ancient surface, with no recent resurfacing by volcanic activity. Its surface is thought to be among the oldest in the solar system, dating back to 4.56 billion years before present. Craters are the only surface features on this moon: It has no mountains, grooves, or scarps. Since its surface is icy, like Ganymede's, its craters also tend to flatten and become palimpsests over time. Among its surface features are two immense impact craters, Valhalla and Asgard (the hall of Odin and the home of the gods, both from Norse mythology). The main depression of Valhalla crater is 190 miles (300 km) in diameter, and its outer rings (formed by the initial shock and collapse of the crater) stretch to 1,900 miles (3,000 km) in diameter. The second basin, Asgard, has outer rings 1,000 miles (1,600 km) in diameter. To the south of Asgard basin lies a region of exceptionally strange landforms. Shown in the figure on page 105, this terrain consists of many sharp ice and dust spires and knobs. As the ice slumps or sublimes, the dark material apparently slides down and collects in low-lying areas. Over time, as the surface continues to erode, the icy knobs will likely disappear, producing a scene similar to the bottom inset. The number of impact craters in the bottom image indicates that erosion has essentially ceased in the dark plains shown in that image, allowing impact craters to persist.

The knobs are about 260 to 330 feet (80 to 100 m) tall, and they may consist of material thrown outward from a major impact billions of years ago. The smallest features discernable in the images are about 10 feet (3 m) across.

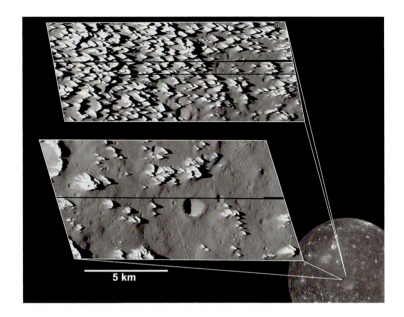

Spires of ice on Callisto are thought to be made by erosion and sublimation of ice from the surface. (NASA/JPL/Galileo)

Callisto is large enough that it should be at least partly differentiated, based on a combination of its size and the coldness of its position in the solar system. Callisto has the lowest density of any of Jupiter's satellites (114 lb/ft³, or 1,800 kg/m³). This very low density implies that the moon consists mainly of ices and contains less rock than some of Jupiter's other moons. Callisto is thought to have an icy crust about 125 miles (200 km) thick, possibly underlain by a salty liquid ocean. The center of the planet is thought to be a combination of ices and rock, with no differentiated iron core.

9. Themisto

Themisto is a small, irregular moon that orbits inside the Himalia prograde group (this group includes Leda, Himalia, Lysithea, Elara, and S/2000 J11) and outside the Galilean satellites, but does not belong to either group. The Galileans all orbit almost perfectly in Jupiter's equatorial plane and have almost circular orbits. The Himalias have orbital eccentricities between 0.1 and 0.25, and they all orbit at about 27 degrees to

(continues on page 108)

RHEOLOGY, OR HOW SOLIDS CAN FLOW

Rheology is the study of how materials deform, and the word is also used to describe the behavior of a specific material, as in "the rheology of ice on Ganymede." Both ice and rock, though they are solids, behave like liquids over long periods of time when they are warm or under pressure. They can both flow without melting, following the same laws of motion that govern fluid flow of liquids or gases, though the timescale is much longer. The key to solid flow is viscosity, the material's resistance to flowing.

Water has a very low viscosity: It takes no time at all to flow under the pull of gravity, as it does in sinks and streams and so on. Air has lower viscosity still. The viscosities of honey and molasses are higher. The higher the viscosity, the slower the flow. Obviously, the viscosities of ice and rock are much higher than those of water and molasses, and so it takes these materials far longer to flow. The viscosity of water at room temperature is about 0.001 Pas (pascal seconds), and the viscosity of honey is about 1,900 Pas. By comparison, the viscosity of window glass at room temperature is about 10^{27} Pas, the viscosity of warm rocks in the Earth's upper mantle is about 10^{19} Pas.

The viscosity of fluids can be measured easily in a laboratory. The liquid being measured is put in a container, and a plate is placed on its surface. The liquid sticks to the bottom of the plate, and when the plate is moved, the liquid is sheared (pulled to the side). Viscosity is literally the relationship between shear stress σ and the rate of deformation ε. Shear stress is pressure in the plane of a surface of the material, like pulling a spatula across the top brownie batter.

$$\eta = \frac{\sigma}{\varepsilon}.$$

The higher the shear stress needed to cause the liquid to deform (flow), the higher the viscosity of the liquid.

The viscosity of different materials changes according to temperature, pressure, and sometimes shear stress. The viscosity of water is lowered by temperature and raised by pressure, but shear stress does not affect it. Honey has a similar viscosity relation with temperature: The hotter the honey, the lower its viscosity. Honey is 200 times less viscous at 160°F (70°C) than it is at 57°F (14°C). For glass, imagine its behavior at the glasshouse. Glass is technically a liquid even at room temperature, because its molecules are not organized into crystals. The flowing glass the glassblower works with is simply the result of high temperatures creating low viscosity. In rock-forming minerals, temperature drastically lowers viscosity, pressure raises it moderately, and shear stress lowers it, as shown in the accompanying figure.

Latex house paint is a good example of a material with shear-stress dependent viscosity. When painting it on with the brush, the brush applies shear stress to the paint, and its viscosity goes down. This allows the paint to be brushed on evenly. As soon as the shear stress is removed, the paint becomes more viscous and resists dripping. This is a material property that the paint companies purposefully give the paint to make it perform better. Materials that flow more easily when under shear stress but then return to a high viscosity when undisturbed are called thixotropic. Some strange materials, called dilatent materials, actually obtain higher viscosity when placed under shear

(continues)

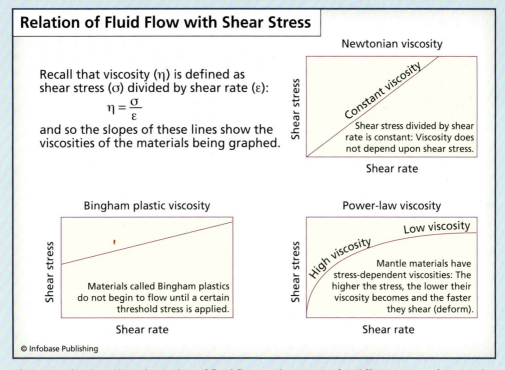

Relation of Fluid Flow with Shear Stress

Recall that viscosity (η) is defined as shear stress (σ) divided by shear rate (ε):

$$\eta = \frac{\sigma}{\varepsilon}$$

and so the slopes of these lines show the viscosities of the materials being graphed.

Newtonian viscosity

Shear stress

Constant viscosity

Shear stress divided by shear rate is constant: Viscosity does not depend upon shear stress.

Shear rate

Bingham plastic viscosity

Shear stress

Materials called Bingham plastics do not begin to flow until a certain threshold stress is applied.

Shear rate

Power-law viscosity

Shear stress

Low viscosity

High viscosity

Mantle materials have stress-dependent viscosities: The higher the stress, the lower their viscosity becomes and the faster they shear (deform).

Shear rate

© Infobase Publishing

These graphs show the relationship of fluid flow to shear stress for different types of materials, showing how viscosity can change in the material with increased shear stress.

(continued)

stress. The most common example of a dilatent material is a mixture of cornstarch and water. This mixture can be poured like a fluid and will flow slowly when left alone, but when pressed it immediately becomes hard, stops flowing, and cracks in a brittle manner. The viscosities of other materials do not change with stress: Their shear rate (flow rate) increases exactly with shear stress, maintaining a constant viscosity.

Temperature is by far the most important control on viscosity. Inside the Earth's upper mantle, where temperatures vary from about 2,000°F (1,100°C) to 2,500°F (1,400°C), the solid rocks are as much as 10 or 20 orders of magnitude less viscous than they are at room temperature. They are still solid, crystalline materials, but given enough time, they can flow like a thick liquid. The mantle flows for a number of reasons. Heating in the planet's interior makes warmer pieces of mantle move upward buoyantly, and parts that have cooled near the surface are denser and sink. The plates are also moving over the surface of the planet, dragging the upper mantle with them (this exerts shear stress on the upper mantle). The mantle flows in response to these and other forces at the rate of about one to four inches per year (2 to 10 cm per year).

Rocks on the planet's surface are much too cold to flow. If placed under pressure, cold, brittle surface rocks will fracture, not flow. Ice and hot rocks can flow because of their viscosities. Fluids flow by molecules sliding past each other, but the flow of solids is more complicated. The individual mineral grains in the mantle may flow by "dislocation creep," in which flaws in the crystals migrate across the crystals and effectively allow the crystal to deform or move slightly. This and other flow mechanisms for solids are called plastic deformations, since the crystals neither return to their original shape nor break.

(continued from page 105)

Jupiter's equatorial plane. Themisto, however, has an eccentricity of 0.248 and orbits at a very high angle of 43 degrees to the equatorial plane. Little is known about Themisto. It is likely to be a captured asteroid, perhaps in a degrading orbit preparing to crash into the planet.

10. Leda

Leda is a tiny moon that orbits at 28 degrees to Jupiter's equator, 6.9 million miles (11 million km) from the surface of Jupiter. Its orbital eccentricity (0.148) and orbital inclination imply that Leda is a captured asteroid. Leda is named after a queen of Sparta in ancient Greek mythology who was seduced by Zeus when he came to her in the form of a swan, and she subsequently gave birth to eggs that hatched into Castor and Pollux. Leda was also the mother of Helen of Troy.

11. Himalia

Himalia is the brightest of Jupiter's outer satellites, though its albedo is just 0.03. Like Leda and many others of Jupiter's moons, Himalia is likely an asteroid that was captured and placed into orbit by Jupiter's gravitational field. Himalia is named after a nymph in Greek mythology who bore three sons of Zeus.

12. Lysithea

Lysithea is another small and poorly known satellite. Lysithea was named for a daughter of Oceanus who was one of Zeus's lovers. Although Lysithea is very small, with a radius of only 11 miles (18 km), it was discovered in 1938, before the more precise techniques of modern astronomy were developed.

Beyond Lysithea are Elara and S/2000 J11, both members of the Himalia prograde irregular group. Outside the Himalia group, the remainder of Jupiter's moons are small, poorly known captured asteroids that orbit in a retrograde sense. The Ananke retrograde group has orbits at 145 to 150 degrees to Jupiter's equatorial plane, and the more distant moons have orbits that are that steep or even steeper.

Jupiter has by far the largest known population of moons of any planet in the solar system, almost twice the number of moons of the second contender, Saturn. No doubt a great many additional moons will be added to Jupiter's list over the coming decade, but it is almost a certainty that all will be small captured asteroids with highly inclined and elliptical orbits, a sort of solar system remainder swarm attracted to the giant gravity well of Jupiter. Jupiter's large satellites, including Ganymede, the largest moon in the solar system, display a mind-boggling variety of compositions, appearances, and behaviors, creating a compelling natural laboratory for the study of planetary evolution. The icy Europa and its near neighbor Io form one of the great contrasts in planetary science: They each have new surfaces, caused in Europa's case by floods of ice from its watery interior, and in Io's by floods of unusually hot magma. While Europa is a candidate location for the development of life, Io is an inhospitable place, covered with sulfur, bombarded by Jupiter's magnetic field, and continuously re-covered by magma in volumes far greater than are produced on Earth each year. These moons will no doubt be the targets of future space missions.

Missions to Jupiter

Jupiter is large enough and bright enough that even early observers were able to learn a substantial amount about the planet. When Galileo first began viewing the sky with his newly made telescope, he almost immediately found the four largest moons, now called the Galilean satellites. He recorded their orbits around Jupiter; this was the first major support for the Copernican theory that planets orbit around the Sun. (In this case, the evidence was simply that the moons orbited around Jupiter, rather than around the Earth, which everything was assumed to orbit.) The Inquisition sentenced Galileo to house arrest for the heresy of a solar system not centered on the Earth, though this was a light sentence, since they would have been within their rights at the time to torture and execute him for his outrageous ideas.

Similarly, the Great Red Spot has been observed from the Earth for over 300 years. It may have been seen first by Giovanni Cassini, the great Italian astronomer, or by Robert Hooke, the English experimentalist. Though the large features of the planet are easily seen through even the smallest telescopes, it took space missions to discover Io's abundant volcanism, many of Jupiter's smaller moons, and the many details of its composition and atmosphere that are now known.

Following are brief descriptions of the seven space missions that have visited Jupiter to date. The excitement created by the first images sent by *Pioneer 10* was intense, and though newer science by more recent missions was perhaps more exciting and revolutionary, people have become more inured to images of distant planets, thinking them more as commonplace pictures than as results of exceptional scientific and engineering endeavors.

Pioneer 10 and 11 1972, 1973
American

Pioneer 10 arrived at Jupiter on December 3, 1973, and *Pioneer 11* arrived December 2, 1974. *Pioneer 10* performed the first flyby of Jupiter by any space mission, at 82,656 miles (132,250 km) above the clouds. *Pioneer 11* brushed by just 26,800 miles (42,900 km) from the cloud tops. This is just one-tenth the distance from the Earth to the Moon, so the missions were able to take high-resolution photos of Jupiter's atmospheric structure, and also to take measurements of the planet's magnetic field.

Voyager 1 and 2 1977
American

After two years in space, the Voyager missions reached Jupiter in 1979, sent by NASA and the Jet Propulsion Laboratory. The *Voyager 1* approach phase began January 4, 1979, with encounter of the Jovian system 60 days later. It passed close to all four Galilean moons. *Voyager 2* encountered the system on July 9, 1979, taking close-up photos of Callisto, Ganymede, Europa, and Amalthea. The excellent photos from the Voyager missions revealed Jupiter's rings, active volcanoes on Io, and the diversity of surfaces on the Jovian moons. *Voyager* data also brought the number of known moons of Jupiter to 16.

Galileo 1989
American

NASA launched *Galileo* on October 18, 1989, seven years after its original launch date, and its mission ended in a blaze of success on September 21, 2003. *Galileo* explored the Jupiter system for seven years, having taken nine years to reach the planet. The mission was redesigned five times, as specifications for the trip changed, and different launch dates required different flight paths through the solar system. The spacecraft was designed to be launched from the space shuttle, but because of delays in the launch date that created problems with the route to the outer solar system, when it was finally launched, *Galileo* did not have enough propellant to take a direct route. An ingenious system of gravitational boosts were used to get the probe all the way to the outer solar system, starting with a launch actually toward the inner solar system, where it swung around Venus, then came back to swing twice around the Earth, before catapulting into the outer solar system. When *Galileo* came back by Earth the second time, a special communication had to come from NASA to the North American Aerospace Defense Command that the fast-moving object seeming to come from the Middle East was not an attack but the return of their own spacecraft on its way to Jupiter.

Two years after launch, *Galileo* had reached the asteroid belt, and was scheduled to open its large antenna and start streaming data back to Earth. When the command was sent, the antenna stuck: It would not open! With only the secondary, smaller antenna, *Galileo* would only be able to send data at a rate that would transmit one picture per month, a completely inadequate rate. What followed was one of the most daring engineering feats of the space age: The original designers of the aged computer processors were brought out of retirement for consultation, and over the course of a year or so, all the software in *Galileo* was replaced. An error could have meant a complete loss of communication with the spacecraft, but the radio link worked, and the software replacement worked, and in the end *Galileo* was able to send 200 images per month. There was still a team assigned to make the large antenna open, and they worked at it for five years, without success.

Galileo took pictures of the four largest Jovian satellites, and studied Jupiter's atmosphere, magnetic field, rings, and small satellites, even dropping a probe into Jupiter's clouds.

The probe entered Jupiter's atmosphere 280 miles (450 km) above the one-bar pressure level and began measuring composition, temperature, and pressure, and sending the data back to *Galileo*. Its first parachute opened at 14 miles (23 km) above the one-bar level, and soon after, the heat shield dropped off. The probe survived to an atmospheric level of 22 bars, when its transmission ended.

Galileo conducted 40 flybys of planets and moons, far more than any other space mission. It passed close by two asteroids, Gaspra and Ida, and took the first high-resolution images of asteroids. In these images it was discovered that Ida has its own orbiting moon! Some scientists refer to the tiny moons of asteroids as "moonlets." *Galileo* was the first to document active volcanism on Io, and it photographed the pieces of the Shoemaker-Levy comet falling into Jupiter. It detected nine new moons of Jupiter.

On September 21, 2003, *Galileo* was directed to fly into Jupiter and destroy itself. This directive was made necessary by discoveries *Galileo* itself had made about the moon Europa. Data from *Galileo* indicates that, with its probable water oceans, Europa is a highly possible site for life in the solar system. Rather than risk contaminating Europa with any microscopic life still on *Galileo,* should *Galileo* crash onto Europa as it aged and its systems failed, NASA made the determination that it should be definitively destroyed in Jupiter's upper atmosphere, where all possible life on it will be completely vaporized, while *Galileo* was still responsive enough to allow guidance from Earth.

Ulysses 1990
American

Ulysses is a joint venture of NASA and the European Space Agency and was launched in 1990. Its primary goal is to observe in the regions of space above the Sun's north and south poles, but that is a particularly difficult orbit to attain. The Earth is moving so swiftly in the plain of the Sun's equator that the mission must go to great lengths to cancel that velocity and create an orbit at right angles to it.

The easiest way to reach a polar orbit is to use a gravity assist from another planet, and it takes a planet with the vast gravity field of Jupiter to meet this requirement. *Ulysses* still had to leave Earth's gravity field at seven miles per second (11.3 km/sec), the fastest interplanetary spacecraft ever launched. The spacecraft reached Jupiter about a year and a half after launch, and it made its closest approach to Jupiter on February 8, 1992. *Ulysses* made detailed and valuable measurements of Jupiter's magnetic field during the 12 days it flew within the magnetosphere. After leaving Jupiter, the craft traveled back to the inner solar system to arrive at the Sun in 1994.

Cassini-Huygens 1997
American and European

Cassini-Huygens was launched in 1997 as a joint effort of NASA, the Jet Propulsion Laboratory, the European Space Agency, and the Italian Space Agency. Seventeen nations contributed to building the spacecraft. After seven years in space and four gravity assists from planets, *Cassini* arrived at Saturn, its main mission goal, in 2004. Along the way, *Cassini* flew by Jupiter while *Galileo* was there, and both craft obtained data at the same time. *Cassini* observed Jupiter from October 1, 2000, to March 31, 2001, with its closest approach on December 30, 2000, at a distance of 9.7 million km.

New Horizons 2006
American

New Horizons launched January 19, 2006. The mission swung past Jupiter for observations and a gravity boost in February 2007. The gravity boost allowed the mission to observe Jupiter for five months. During that period, the spacecraft saw the development of a new storm, called the Little Red Spot, which was later destroyed in an encounter between the Great Red Spot and a smaller storm called Oval BA. *New Horizons* also photographed huge volcanic eruptions from the moon Io's volcano Tvashtar.

New Horizons will reach Pluto and Charon in 2015. The spacecraft will then head deeper into the Kuiper belt to study one or more of the bodies in that vast region, at least a billion miles beyond Neptune's orbit. *New Horizons* has many science partners, including the Johns Hopkins University Applied Physics Laboratory, the Southwest Research Institute, Ball Aerospace Corporation, Stanford University, and NASA. The mission includes instrumentation for visible-wavelength surface mapping and infrared and ultraviolet imaging to study surface composition and atmosphere, radiometry, and solar wind measurements.

Missions to Jupiter have been planned, canceled, reinstated, and canceled again. One exciting mission was the Jupiter Icy Moons Orbiter, but its rocky path to a green light ended instead by cancellation by Congress. The Juno New Frontiers mission was briefly canceled and is now reinstated, with a planned launch date of August 5, 2011. The variety and especially the potential for the development of life on Jupiter's icy moons will continue to draw the attention of the space community, and future missions to the moons of Jupiter are almost inevitable.

PART TWO

SATURN

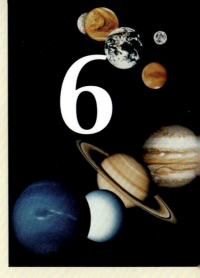

Saturn: Fast Facts about a Planet in Orbit

Saturn is the most remote of the planets known to ancient man. Though its unique and immense rings have made it a sort of symbol for all planets, these rings were undiscovered until the mid-17th century (see the figure on page 120). The rings and moons themselves form a sort of miniature replica of the early solar system, with interactions among dust and particles and larger bodies all orbiting together within the same magnetic field. Now the complex interactions between its hundreds of rings and dozens of moons are the subject of intense research, especially since the *Cassini-Huygens* spacecraft entered Saturn's orbit on June 30, 2004.

For the first time, Saturn is being studied with the same intensity that has been devoted to other planets. The *Cassini-Huygens* mission has been delivering enough data and images to Earth to enable scientists to answer some of the outstanding questions about this spectacular planet. Scientists hope that *Cassini-Huygens* will ultimately provide answers to questions about the internal structure of the planet, interactions among its magnetic field, moons, and rings, and most particularly, questions about its giant moon, Titan, and its icy moon Enceladus.

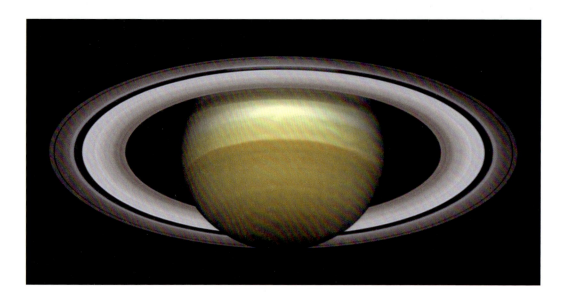

Saturn's gorgeous rings have made it a symbol for all planets. Though the planet has been known since prehistory, its rings could not be seen until the invention of the telescope. (NASA and the Hubble Heritage Team [STScI/AURA] R. G. French [Wellesley College], J. Cuzzi [NASA/Ames], L. Dones [SwRI], and J. Lissauer [NASA/Ames])

Titan was the last object in the solar system with a mysterious surface: Its thick atmosphere prevented the two Voyager missions from seeing it, and the best images of Titan's surface are tantalizing blurry images from the *Hubble Space Telescope* using special technology. These images hinted that Titan may have oceans and continents. Scientists hypothesized that if Titan did have oceans, they would consist of hydrocarbons and not water, which is frozen harder than rock at the temperatures of Titan's surface. Nonetheless, Titan may be similar in some ways to conditions on the early Earth and may form a laboratory of sorts to answer questions about early life. Titan's atmosphere is thicker than the Earth's but consists primarily of nitrogen, as does the Earth's. Titan's cold temperatures may have preserved it in the conditions of the distant past, similar to Earth before the development of life.

Cassini dropped the *Huygens* probe on December 25, 2004, and it passed through Titan's atmosphere and landed on its surface on January 14, 2005. The measurements and images the probe sent back, along with later data from *Cassini*, confirmed some aspects of Titan: Its surface is covered with drainage gullies similar to Earth's, and it has small lakes dotting its polar regions. The stones on its surface are indeed made of frozen ices. Research continues actively on information *Cassini*

FUNDAMENTAL INFORMATION ABOUT SATURN

While Saturn's diameter is more than nine times as large as the Earth's, it is the least dense of solar system objects. While Jupiter is on average about one-third more dense than water, Saturn is actually less dense than water and would float if it were possible to place it in a bath large enough. The low density of Saturn has another effect: Though the planet's radius is almost 10 times as great as the Earth's, its gravitational acceleration at a pressure equivalent to Earth's surface is very similar to the Earth's. Were it possible to stand on Saturn, the astronaut would feel as heavy as on Earth.

FUNDAMENTAL FACTS ABOUT SATURN	
equatorial radius at the height where atmospheric pressure is one bar	37,450 miles (60,268 km)
polar radius	33,780 miles (54,362 km)
ellipticity	0.098, meaning the planet's equator is almost 10 percent longer than its polar radius
volume	1.98×10^{14} cubic miles (8.27×10^{14} km^3), or 755 times Earth's volume
mass	1.25×10^{26} pounds (5.69×10^{26} kg), or 95 times Earth's mass
average density	43.7 pounds per cubic foot (700 kg/m^3), less than the density of water
acceleration of gravity at the equator at the height where atmospheric pressure is one bar	34.24 feet per second squared (10.44 m/sec^2), or 1.065 times Earth's gravity
magnetic field strength at the surface	2×10^{-5} tesla
rings	hundreds
moons	60 presently known

continues to send back about this body, possibly so similar to ancient Earth.

Each planet and some other bodies in the solar system (the Sun and certain asteroids) have been given their own symbols as shorthand in scientific writing. The symbol for Saturn is shown below.

On average, Jupiter is 5.2 AU from the Sun, that is, 5.2 times farther than the Earth. Saturn, though it is spoken of as the twin of Jupiter because of their many similarities, is almost twice as far from the Sun as Jupiter is. Saturn averages 9.54 AU from the Sun, immensely farther away than Jupiter. That and Saturn's other orbital characteristics are listed in the table on page 123.

Seasons are caused largely by the tilt of the planet's rotational axis, called its obliquity (see figure on page 124). As a planet rotates around the Sun, its axis always points in the same direction. (The axis does wobble slightly, a movement called *precession*.) The planet with the most extreme obliquity is Venus, with an obliquity of 177.3 degrees, followed by the dwarf planet Pluto, with an obliquity of 122.53 degrees. These obliquities above 90 degrees mean that the planet's north pole

Many solar system objects have simple symbols; this is the symbol for Saturn.

Symbol for Saturn

© Infobase Publishing

ORBITAL PARAMETERS FOR SATURN

rotation on its axis ("day")	10 hours, 14 minutes at the equator, and 10 hours, 40 minutes at the poles; the second-shortest day length in the solar system after Jupiter
rotation speed at equator	22,081 MPH (35,536 km/hour)
rotation direction	prograde (counterclockwise when viewed from above the north pole)
sidereal period ("year")	29.46 Earth years
orbital velocity (average)	6.01 miles per second (9.67 km/sec)
sunlight travel time (average)	1 hour, 19 minutes, and 20 seconds to reach Saturn
average distance from the Sun	885,904,700 miles (1,426,725,400 km), or 9.54 AU
perihelion	885,519,000 miles (1,349,467,000 km), or 9.021 AU from the Sun
aphelion	934,530,000 miles (1,503,983,000 km), or 10.054 AU from the Sun
orbital eccentricity	0.0542
orbital inclination to the ecliptic	2.48 degrees
obliquity (inclination of equator to orbit)	26.73 degrees

has passed through its orbital plane and now points south. This is similar to Uranus, which has a rotational axis tipped until it almost lies flat in its orbital plane. Some scientists also think that Mercury's obliquity is 180 degrees, not 0 degrees, as usually reported. Earth's obliquity is 23.45 degrees, Mars's is similar, at 25.2 degrees, Jupiter's is 3.12 degrees, Saturn's is 26.7 degrees, and Neptune's is 29.56 degrees. With the exceptions of Mercury and Jupiter, therefore, all the planets have significant seasons caused by obliquity.

When a planet with obliquity has its north pole tipped toward the Sun, the northern hemisphere receives more direct sunlight than the southern hemisphere does. The northern

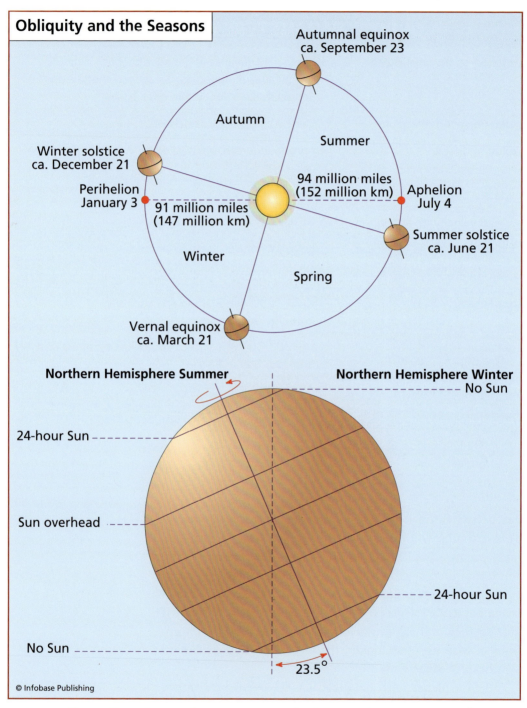

Obliquity and the Seasons

Autumnal equinox
ca. September 23

Autumn

Summer

Winter solstice
ca. December 21

94 million miles
(152 million km)

Aphelion
July 4

Perihelion
January 3

91 million miles
(147 million km)

Summer solstice
ca. June 21

Winter

Spring

Vernal equinox
ca. March 21

Northern Hemisphere Summer

Northern Hemisphere Winter

No Sun

24-hour Sun

Sun overhead

24-hour Sun

No Sun

23.5°

© Infobase Publishing

A planet's obliquity (the inclination of its equator to its orbital plane) is the primary cause of seasons.

hemisphere then experiences summer, and the southern hemisphere is in winter. As the planet progresses in its orbit, revolving around the Sun until it has moved 180 degrees, then the southern hemisphere gets more direct sunlight, and the northern hemisphere is in winter. The more oblique the rotation axis, the more severe the seasons: The hemisphere in summer receives even more sunlight, the other hemisphere even less. Summers are hotter and winters are colder. The obliquity of a planet may change over time, as well. Mars's obliquity may oscillate by as much as 20 degrees over time, creating seasons that are much more extreme. The Moon's stabilizing influence on the Earth has prevented large changes in obliquity and helped maintain a more constant climate, allowing life to continue and flourish.

Summer occurs when the Sun shines most directly on the hemisphere in question, and the intensity of summer must depend on when it occurs in the planet's orbit. If the planet's axis tilts such that the hemisphere has summer at perihelion, when the planet is closest to the Sun, then it will be a much hotter summer than if that hemisphere had summer at aphelion, when the planet is farthest from the Sun (summer at the aphelion will also be shorter, since the planet is moving faster). Planets' orbits do precess, that is, wobble, and so the positions of the seasons change over tens of thousands of years. This leads to a long-term cycle in seasonal severity.

In the northern hemisphere, the midpoint of summer occurs when the north pole points most directly toward the Sun. This is called the summer solstice: the longest day of the year, after which days become shorter. The northern hemisphere's winter solstice, when the south pole points most directly toward the Sun, is its shortest day of the year. This same day is the southern hemisphere's summer solstice, its longest day of the year. The planet's obliquity has the largest control over the severity of seasons (there are secondary effects that also influence the temperature differences between seasons). In the summer the Sun is higher in the sky, and so spends more time crossing the sky, and therefore the days are longer. This gives the Sun more time to heat the planet. In the winter the Sun is lower, and the days are short, giving the Sun less time to heat the planet. Along with summer solstice and winter solstice, there are two other days that divide the year into quarters. The vernal equinox is the day in spring when day and night are the same length (equinox means equal night). The autumnal equinox is the day in fall when day and night are the same length.

Saturn's obliquity has been obvious for centuries from the tilt of its rings. Now, sensitive images of its weather patterns show that the planet may have storms that track the seasons created by the planet's obliquity and orbit around the Sun. Large bright storms appear and disappear on a 30-year cycle, and smaller features can now be tracked as well. Saturn's obliquity is similar to the Earth's and to Mars's, but far higher than its neighbor and most similar planet Jupiter. Because of its obliquity, Saturn may have cycles of weather and atmospheric circulation entirely different from Jupiter's.

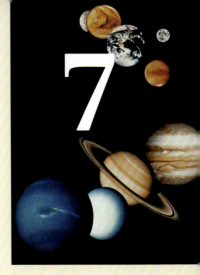

Saturn's Interior: Composition and Magnetic Field

The interior composition, structure, and dynamics of Saturn and Jupiter appear to be remarkably similar, though the internal structures of both planets remain the subjects of hypothesis and speculation, with little firmly established. Saturn does have more helium in its outer layers than does Jupiter, and Jupiter has a stronger magnetic field, but in many ways the planets are twins.

STRUCTURE, TEMPERATURE, PRESSURE

Saturn consists almost entirely of helium and hydrogen. Gas giant planets are thought to have accreted out of the solar nebula through collisions as a core of ice with a small component of rock, up to about 10 Earth masses, at which point the gas of the solar nebula began to be attracted to the protoplanets. Later differentiation may consist mainly of separation of helium and hydrogen.

Knowing that the planet must consist almost entirely of helium and hydrogen based on its density and the abundance of those elements in the solar system, models for the planet's interior can be made based on laboratory physics. The behavior of those elements can be predicted as pressure in the planet increases with depth. In the planet's interior, helium and hydrogen gas gradually become liquid with increasing pressure and depth. At a pressure of about 2.5 million atmospheres (250 GPa), hydrogen changes abruptly into a metallic state. In its metallic state the hydrogen atom's electrons flow freely among the nuclei, creating an electrically conductive region. Beneath the metallic zone probably lies a planet's dense ice-silicate core, thought to make up about one-third of its total mass and reach to about one-quarter of the planet's radius. Saturn's interior is thought to reach a pressure of 45 million atmospheres (4,500 GPa) and a temperature of about 22,000°F (12,000°C). Compare this to the Earth, which reaches only 8,700°F (4,800°C) and 3.5 million atmospheres (360 GPa) in the inner core.

The density of Saturn is low, consisting as it does mainly of helium and hydrogen. The shallow parts of the planet's temperature profile can be measured through stellar *occultation:* As a star's light is eclipsed behind Saturn, its light is extinguished gradually by the increasing density of Saturn's atmosphere. The rate of extinction of the star's light yields information on the density of the atmosphere, which in turn can be related to temperature (the higher the temperature, the less dense the atmosphere). Spectra obtained from a known constituent in the atmosphere, such as methane, can also give temperature information.

INTERNAL HEAT PRODUCTION

Saturn, along with Jupiter, radiates more energy than it receives from the Sun. Saturn's heat budget is not as extreme as Jupiter's, though; some measurements from *Pioneer* and *Voyager* indicated that Saturn receives almost as much energy from the Sun as it produces internally. More recent and exact measurements show that Saturn produces 82 percent more heat internally than it receives from the Sun. The degree of internal energy required to match the solar input is not as excessive as it would be for the Earth: Saturn receives only 1 percent of the solar flux that the Earth receives.

Excess internal energy may come from compositional differentiation or from internal gravitational collapse. In the case of internal compositional differentiation, scientists hypothesize that a fractionation of helium from hydrogen may be the driving force. At pressures of 3 to 5 million bars inside Saturn, hydrogen changes to a metallic state, where atoms share electrons as if they were a gas among the atomic nuclei. When hydrogen reaches this metallic state, helium can no longer mix homogeneously with the hydrogen, and the helium condenses into a liquid and drops further into Saturn's interior as helium rain. The dropping helium loses potential energy as it falls into the gravity field of the planet and releases heat. This kind of falling may continue until material reaches the core of the planet, which may be as hot as 21,600°F (12,000°C). Saturn may also be collapsing gradually under its own gravitational field. Contraction similarly releases heat from both the loss of potential energy and from friction, and so collapse may add to the internal heat budget of the planet.

Saturn and Jupiter have similar radii (37,450 miles, or 60,268 km, for Saturn versus 44,424 miles, or 71,492 km, for Jupiter), though Saturn is considerably less massive than Jupiter (1.25×10^{26} pounds, or 5.69×10^{26} kg, for Saturn versus 4.2×10^{27} pounds, or 1.9×10^{27} kg, for Jupiter). The similarity in radii can be explained if helium differentiation took place much earlier on Jupiter than it did on Saturn. Jupiter may thus hold the majority of its mass in a dense state near its core, while on Saturn, helium is still differentiating and sinking deeper into the planet. Unfortunately, this theory is not supported by helium measurements in the outer portions of these planets: Saturn has a lower helium to hydrogen ratio than Jupiter, implying that less of its helium remains in the outer portions of the planet.

MAGNETIC FIELD

Saturn's magnetic field is 35 times weaker than Jupiter's, but it is still 550 times greater than Earth's. Saturn's magnetic pole and its geographic pole (the pole of its rotation) are aligned to within one degree. As shown in the figure on page 131, the planet is surrounded by a magnetosphere, the volume dominated by the planet's own magnetic field. The solar wind strikes the sunward side of the magnetosphere and compresses, creating a shock wave called the bow shock. The bow shock wraps smoothly around the planet, streaming off the far side in an apron. A region called the magnetopause separates the compressed, heated, smoothly flowing solar wind from the oddly shaped magnetosphere. Finally, a thin region called the magnetosheath separates the bow shock from the magnetopause. On the nightside of the planet the solar wind pulls the magnetosphere into a magnetotail, an extension of the field that can be 25 to 100 times the radius of Saturn in length. In its basic structure, Saturn's magnetic structure as described here is similar to that of any of the planets with their own magnetic field. The size of the planet's field and the varying strength of the solar wind determine the size and the shape of each magnetosphere and its attendant structures.

Like Jupiter's, the polarity of Saturn's field is opposite that of the Earth's, that is, compasses on Saturn would point to its south pole. Saturn's magnetic field lines pass into its south pole and out of its north pole. Saturn's field and Jupiter's field are thought to be produced in the same manner, by convecting currents in its liquid hydrogen layer. The planets also have in common the complexity of their fields: Though both fields are dominated by the dipole component (analogous to a simple bar magnet), there are more complex fields present as well, including quadrupole fields (two positive and two negative poles) and octupole fields (four of each). Unlike any other planet in the solar system, Saturn's magnetic field is almost perfectly aligned with its rotation axis. Since the Earth's field is known to move measurably within human history, the other planets' fields may also be assumed to move. It may simply be a fortuitous moment in the age of the solar system that space observations have caught Saturn's field in this position.

The 15 innermost moons of Saturn orbit either in or through the edges of the planet's magnetosphere. Titan, the outermost of these moons, moves in and out of the magnetosphere depending on its position in its orbit and the strength

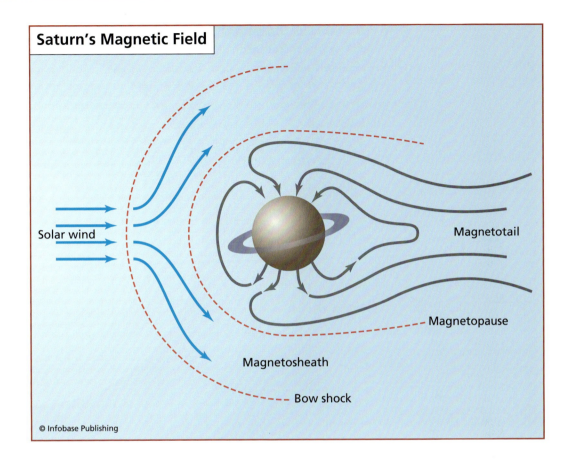

Saturn's Magnetic Field

Solar wind

Magnetotail

Magnetopause

Magnetosheath

Bow shock

© Infobase Publishing

of the solar wind. Titan, Enceladus, Tethys, and Dione all lose gas to Saturn's magnetosphere, adding ions that spiral along the magnetic field lines.

Saturn's aurora is also unique in observations of the solar system. When ions from the solar wind or from nearby moons follow magnetic field lines toward the poles of the planet and strike the upper atmosphere, the energy of the accelerated particles is released as light. All other planetary auroras observed have been bright, colored displays like ours on Earth. Saturn, on the other hand, displays both bright and dark auroras. Its north pole has a dark halo caused by the aurora. The process that produces the aurora may also produce unusual atmospheric particles, and these particles may absorb sunlight more efficiently than the rest of the atmosphere, and this excess absorption may explain the dark aurora.

Saturn's magnetic field, like those of all the planets, is pulled out into a long tail by the force of the solar wind. (Compare to the undistorted shapes of magnetic fields shown in the figure on page 33.)

The model for Saturn's interior is relatively simple, because there is so little direct data. As presented here the model cannot explain the density or heat flow of the planet. Much more remains to be discovered. Somewhat more is known about Saturn's shallow surface movements, and far more is known about its moons.

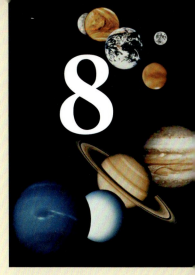

Saturn's Surface Appearance and Conditions

Though Saturn's weather patterns lack Jupiter's intense coloration, the planet still has violently fast winds, giant convective storms, and streaming ribbons of clouds. These weather patterns can be directly observed, unlike the planet's interior processes, but still little is known about them. Saturn's average temperature is -288°F (-178°C) at a pressure of about one bar. The planet's exceptionally cold atmosphere means that ice clouds form at low pressures, and liquid droplet clouds can only exist at greater pressures and hence higher temperatures.

Saturn consists of helium and hydrogen almost exclusively: Only 0.2 percent of its atmosphere by volume is heavier than helium. These heavier constituents create most of the clouds and color of the planet, however. Saturn's clouds are thought to consist mainly of ammonia ice (NH_3), ammonium hydrosulfide (NH_4SH), and water (H_2O) in order of descending altitude. A few even rarer atmospheric components consisting of phosphorus, arsenic, and germanium each combined with hydrogen have been detected in infrared spectra from the planet. The relative abundances of molecules in Saturn's atmosphere are given in the table on page 34.

The upper clouds of ammonium compounds block the water clouds from observer's sight, but the cloud zone is thin and soon changes with depth to clear atmosphere of mainly helium and hydrogen. Aside from the thin layer of colored clouds that decorate the outermost shell of Saturn, there is also a hydrocarbon haze around the planet's poles, possibly caused by electrical interactions from the auroras with the upper atmosphere.

Saturn's atmosphere can be divided into two parts: the first, the troposphere, reaching from the planet's interior with temperature decreasing with height to a point called the tropopause; the second, the stratosphere, beginning at the tropopause and continuing upward with increasing temperature until the atmosphere merges into the vacuum of space. On Saturn the tropopause occurs at a pressure between 60 and 100 millibars and a temperature of -320°F (-193°C). Temperature increases in the stratosphere because of solar heating, and since each higher layer is hotter than its predecessor, the lay-

ABUNDANCE OF SELECTED MOLECULES IN SATURN'S ATMOSPHERE

Molecule	Composition	Number of molecules per molecule of hydrogen
hydrogen	H_2	1
helium	He	0.035
methane	CH_4	0.0045
ammonia	NH_3	0.0003
ethane	C_2H_6	0.000003
phosphine	PH_3	0.0000045
arsine	AsH_3	0.000000003
carbon monoxide	CO	0.0000000016
germane	GeH_4	0.0000000005

ers are stable to mixing: Higher temperatures create less dense material, so the higher material is buoyant with respect to all that lies beneath it. Because in the tropopause the temperature gradient is reversed, the troposphere is unstably stratified with respect to density, and it is prone to convection and creation of storms.

As with the other gas giant planets, Saturn's uppermost stratosphere, called the exosphere, is exceptionally hot. Its temperature cannot be explained by solar heating alone. The unusual heat may be caused by charged particles moving through Saturn's magnetic field that deposit energy in the uppermost atmosphere, or perhaps by electric currents dissipating in the upper atmosphere. Even gravity waves from deeper in Saturn have been proposed as the source of this energy. There is no single leading theory for this phenomenon at present. Though Saturn's clouds form spots and light and dark bands in parallel with its equator, these markings are much less well-defined than on Jupiter. Saturn has an especially intense equatorial jet with winds as fast as 1,100 MPH (1,800 km/hr) in the direction of the planet's rotation. This equatorial jet is four times as fast as Jupiter's. The high winds are not understood: On Jupiter, the *Galileo* probe measured strong winds down to pressures as high as 20 bars, which is well below the depths to which solar energy reaches. The winds cannot therefore be caused by solar heating alone.

Saturn's cloud surface has many visible patterns, including discernible storms similar to Jupiter's White Oval Spots. The *Voyager 2* mission observed anticyclonic brown ovals in Saturn's northern hemisphere that were 3,100 miles (5,000 km) × 1,800 miles (3,000 km) in extent. Large storms on Saturn have been observed to appear and disappear over time since the 19th century, when telescopes were reliably good enough to discern individual storms with certainty. Brilliant white storms were seen by various astronomers in 1876, 1903, 1933, 1960, and 1990, establishing a 30-year cycle for their appearance. The appearance and disappearance of giant storms seems to be tied to Saturn's seasons, and thus to be driven by solar heating. They appear between about five and 60 degrees

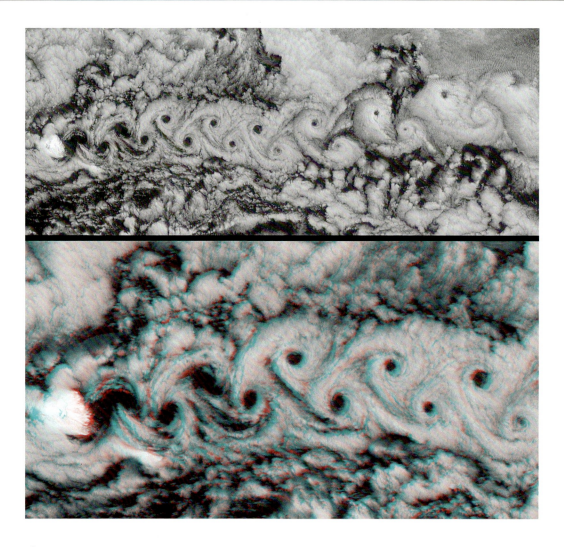

This vortex stream is in a cloud on Earth that is moving rapidly past an island in the north Atlantic, forming structures strikingly similar to those on Saturn. (NASA/GSFC/ LaRC/JPL, MISR Team)

latitude, where solar heating is at a maximum, and are thought to consist of water clouds, something like thunderheads thousands of miles across. Ribbon structures in Saturn's upper atmosphere can be seen in some images from *Voyager 2*. Scientists think the ribbons are waves within a rapidly moving jet stream. These ribbons shed vortices to their sides that appear something like the vortices that shed from the sides of the trail of a fast airplane in a cloud, or that are formed in clouds moving rapidly past an island in the ocean (see figure on this page). The forms of the ribbons and the vortices indicate that the ribbons are moving at about 335 MPH (540 km/hr).

Individual spots, mainly discrete storms, move with the flow of the latitude in which they form. *Voyager* observations in 1980 and 1981 were the first to accurately track the movement of these spots, and they provided the first measurement of Saturn's *differential rotation:* The planet's equator rotates faster than its poles, a phenomenon observed on all the gas planets and the Sun. Though atmospheric features are clearly visible in the image shown on page 120, compared to Jupiter, Saturn's visible surface has a relatively uniform color. Its uniformity is an ongoing mystery: Are many of the convective cells (cyclonic or anticyclonic storms) made of the same material? The material on Saturn may in fact be more uniform, but clouds on Saturn are also located deeper in the atmosphere than they are on Jupiter, because Saturn is in general colder at any given pressure than is Jupiter. Cloud patterns on Saturn are therefore more obscured by overlying hazes than they are on Jupiter.

Saturn's upper atmosphere is directly visible from the Earth and from space missions, but its observation has created more questions than answers. The polar hydrocarbon haze, exceptionally hot exosphere, the depth of fast winds, and the uniformity of cloud color are all persistent questions without clear answers.

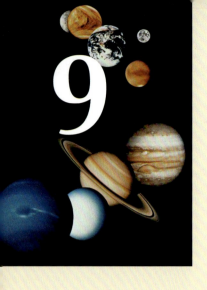

9

Saturn's Rings and Moons

Rings and moons are related in complex ways. Rings are thought to be transitory structures in solar system history, probably lasting no more than hundreds of millions of years. Moons, by contrast, can be stable over the age of the solar system, or they can be unstable and either break up under the gravitations forces of the planet or even crash into the planet itself. The formation of a giant set of rings such as Saturn's is not well understood. An existing moon may have broken up under the force of an impact, or a passing icy body may have been trapped in Saturn's gravitational field and broken. Now that they do exist, though, the rings and moons affect one another in complex ways. Shown in the photo at right, some of Saturn's rings are shaped and maintained by the gravitational fields of nearby moons, and other rings are created from particles shed from moons.

RINGS

In 1610 Galileo observed Saturn's rings as strange hoops seemingly attached to the edges of the planet. His telescope was not sufficiently developed to allow him to see that the rings were continuous (it only magnified by a factor of 20), and, as a

result, they made no physical sense to him. He speculated that Saturn had twin moons orbiting it very closely. Much to his surprise, in 1612 the hoops had vanished: Saturn was aligned so that the edge of the ring plane pointed exactly at the Earth, and due to their extreme thinness the rings effectively disappeared. He wrote: "I do not know what to say in a case so surprising." In 1616 the rings were back, looking more like elliptical hoops than ever. Galileo knew a planet could not have handles like a cup, since planetary spin requires that the planet be largely symmetric around it axis of rotation.

In 1655 Christiaan Huygens, an astronomer from a prominent Dutch diplomatic family, saw the rings through the 50-power telescope that he had designed himself and surmised what they are: a thin flat ring around the planet, nowhere touching the planet itself. Huygens was able to have instruments and make connections unavailable to other scientists of his time from less prominent families. Huygens was tutored in mathematics by René Descartes, one of the greatest mathematicians and philosophers. (He invented Cartesian geometry and made the famous statement "I think, therefore I am.") Along with the mathematical luminaries Pierre de Fermat and Blaise Pascal, Christiaan Huygens corresponded regularly with Marin Mersenne, the mathematician who effectively created the theory of prime numbers that is used today. Other scientists speculated in more wild ways about the rings, but Huygens quietly continued to study the planet. Though he had discovered the rings in 1655, he had been involved with publications concerning his invention of the pendulum clock and was not prepared to write a treatise on Saturn at that moment. In the strangely beautiful convention of the day, Huygens published his discovery as an anagram, which though virtually undecipherable at the time, would still prove the primacy

Cassini imaged Saturn's most prominent feature in natural color visible light; gaps, gravitational resonances, and wave patterns are all present. (NASA/JPL/ Space Science Institute)

of his discovery should someone else see the rings before he could publish properly. The anagram Huygens published was

VVVVVVV.CCC.RR.HMBQX
aaaaaaa,ccccc,d,eeeee,g,h, iiiiiii,llll,mm,nnnnnnnnn,oooo,pp,q,
rr,s,ttttt,uuuuu

Unscrambled, it forms, *"Annulo cingitur, tenui, plano, nuspua cohaerente, ad ecliptican inclinato,"* or in English: "He is surrounded by a thin flat ring, which does not touch him anywhere and is inclined to the ecliptic."

In 1659 Huygens finally published his book *Systema Saturnium,* containing his ring observations and theories that every 14 or 15 years the Earth passes through the plane of the rings, making them invisible from the Earth's vantage point. Huygens's detailed descriptions and explanations for ring observations gradually won over the scientists of the day; his preeminence in lens grinding had allowed him to precede his contemporaries because of better instruments. Still, in 1659 almost all scientists believed that Saturn's rings were solid. At one time, only Giovanni Cassini and Jean Chapelain (a minor poet oddly aligned with the French Académie des Sciences) were clearly in favor of a model in which the rings consisted of multitudes of tiny satellites.

When Saturn's rings are seen from the unlit side the dense rings appear black because they block the light, while thin rings allow light to pass through. (NASA/JPL/ Voyager)

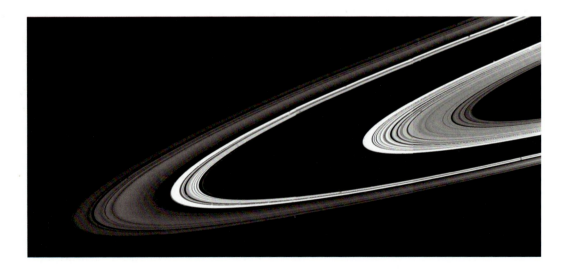

In 1656 Cassini first saw a gap in the rings, now named Cassini's Division, forming the boundary between the A and B rings. In 1850 the C ring was found. In 1825 Henry Kater, an English physicist, reported seeing three gaps in the A ring, but for decades no one else could resolve these features. Finally, in 1837 Johann Franz Encke, a German astronomer who was trained by the Karl Friedrich Gauss, observed a dark band in the A ring that corresponded to one of Kater's gaps, and though Encke did not resolve this band as a gap, it came to be known as Encke's Division. Encke's Division was not clearly seen in a telescope until 1888, when James Keeler, an American astronomer working at the Lick Observatory, obtained a good view. In 1850 William and George Bond, father-and-son American astronomers who developed the technique of photography through a telescope, observed a narrow dark ring interior to the B ring. This new dark ring was originally named the Crepe Ring but became known as the C ring.

Cassini's Division can be seen clearly in the unusual image, shown above, of Saturn's rings taken from the unlit side. When seen with light shining on their far side, relatively opaque areas like the B ring turn black, while lightly populated zones, such as the C ring and the Cassini Division, allow diffuse sunlight to pass through and so appear bright. The A ring, with intermediate opacity, is at an intermediate level of brightness.

George Bond concluded that three solid rings could not be stable in orbit around a planet, and decided the rings must be fluid. Other observers at about the same time were able to see Saturn through the C ring, certainly supporting the idea that the ring could not be solid. In 1849 Edouard Roche, a French physicist, had proposed that a satellite had approached Saturn too closely and had been torn apart by tidal forces, forming the rings from fluid droplets of this destroyed satellite. (For more, see the sidebar "Why Are There Rings?" on page 144.) Roche noted that the radius around Saturn interior at which large satellites would be torn apart by tidal forces was just external to the rings themselves. Roche's excellent calculations, which stand today, did not convince the majority of his contemporaries. Two centuries were required for the basic structure of the rings to be deciphered.

The solid ring theory prevailed, as sometimes an entrenched theory in science does when it is defended by established scientists in the face of newer researchers. In the mid-18th century, Pierre-Simon de Laplace, an exceptional French mathematician who specialized in celestial mechanics, and William Herschel, the German-born English discoverer of Uranus, both wrote that they believed Saturn was circled by two solid planar rings, continuing the majority view from the previous century. Finally, in 1859, James Clerk Maxwell, an eminent English mathematical physicist whose achievements are perhaps second only to Newton's, stated that Saturn's rings must consist of an "indefinite number of unconnected particles," definitively ushering in the now-proven view of ring structure. Maxwell's theory was proven in 1895, when James Keeler and William Campbell, then Keeler's assistant but later director of the Lick Observatory himself, were able to observe that the inner parts of the rings orbit more rapidly than the outer parts, demonstrating that they could not be solid.

The D and E rings were not seen until the 1960s, well into the era of giant institutional observatories. Spectra taken from Earth in 1970 showed the rings to consist mainly of water ice, a result that came even before *Pioneer* was launched in 1973. Thus, the A, B, C, D, E, and F rings were all discovered from ground-based observations, and all subsequent major ring discoveries to

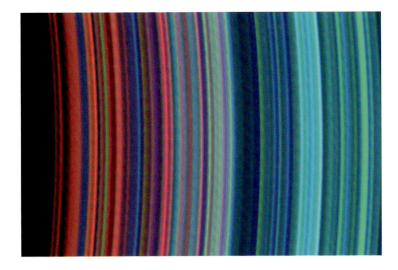

From left to right, this image shows the outer portion of Saturn's C ring and inner portion of the B ring. The B ring begins a little more than halfway across the image. (NASA/JPL/University of Colorado)

date have been made by space missions. Saturn's gorgeous outer B and inner C rings can be seen in the photo at right.

When *Voyager 1* reached Saturn in 1980, its images revealed the tiny F ring that had been proposed in 1979 by Peter Goldreich, now a professor at the California Institute of Technology, and Scott Tremaine, now a professor at Princeton University, when they theorized that two moons could shepherd such a thin ring. The Voyager missions also revealed the existence of the G ring, as well as spokes in the B ring, braiding in the F ring, and the new ring gaps named Maxwell, Huygens, and Keeler.

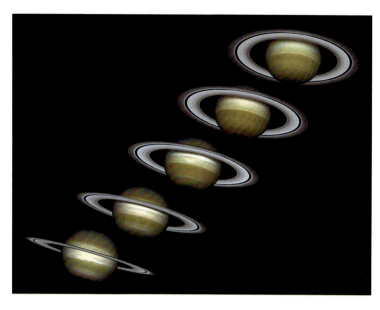

Between 1996 and 2000, Saturn moved relative to the Earth such that astronomers could see the rings first edge-on, and then gradually more and more fully from their bottoms, as seen in the images below from the *Hubble Space Telescope,* captured as Saturn moved from autumn toward winter in its northern hemisphere. Saturn's obliquity, 27 degrees, is similar to the Earth's 23-degree tilt. The first image in this sequence, on the lower left, was taken soon after the autumnal equinox in Saturn's northern hemisphere. By the final image in the sequence, on the upper right, the tilt is nearing its extreme, or winter solstice in the northern hemisphere. Ring crossings are particularly fortuitous times to search for new satellites, since the view of the space around Saturn is obscured neither by the ring material itself nor by the brightness of reflected light.

Saturn's bright rings are less than three miles (2 km) thick but 150,000 miles (250,000 km) in radius, equivalent to about two-thirds the distance from the Earth to the Moon. The

The Hubble Space Telescope *took a series of images showing Saturn's changing seasons. Between 1996 and 2000, Saturn moved relative to the Earth such that astronomers could see the rings first edge-on and then gradually more and more fully from beneath. (NASA and the Hubble Heritage Team [STScI/AURA], R. G. French [Wellesley College], J. Cuzzi [NASA/Ames], L. Dones [SwRI], and J. Lissauer [NASA/Ames])*

(continues on page 146)

WHY ARE THERE RINGS?

Galileo Galilei first saw Saturn's rings in 1610, though he thought of them as handles on the sides of the planet rather than planet-encircling rings. After this first half observation, there was a hiatus in the discovery of planetary ring systems that lasted for three and a half centuries, until 1977, when Jim Elliot, an astronomer at the Massachusetts Institute of Technology and the Lowell Observatory, saw the blinking of a star's light as Uranus passed in front of it and correctly theorized that Uranus had rings around it that were blocking the light of the star. Two years later, *Voyager 1* took pictures of Jupiter's rings, and then in 1984, Earth-based observations found partial rings around Neptune. Now it is even hypothesized that Mars may have very tenuous rings, with an optical depth of more than 10^{-8} (meaning that almost all the light that shines on the ring goes straight through, without being scattered or reflected).

There are two basic categories of planetary rings. The first involves rings that are dense enough that only a small percentage of the light that shines on them passes through. These dense rings are made of large particles with radii that range from centimeters to meters. Examples of these dense rings are Saturn's main rings A and B and the Uranian rings. The second involves tenuous rings of particles the size of fine dust, just microns across. In these faint rings the particles are far apart, and almost all the light that strikes the ring passes through. Jupiter and Saturn's outermost rings are of this faint type. Neptune's rings, however, do not fall into either neat category.

In dense rings the constant collisions between particles act to spread out the rings. Particles near the planet lose speed when they collide with other particles, and thus fall closer to the planet. Particles at the outer edges of the ring tend to gain speed when they collide, and so move farther from the planet. In a complex way, the changes in velocity and redistributions of angular momentum act to make the ring thinner and thinner in depth while becoming more and more broadly spread from the planet.

Most dense rings exist within a certain distance from their planet, a distance called the Roche limit. Within the Roche limit, the tidal stresses from the planet's gravity overcome the tendency for particles to accrete into bodies: The gravitational stresses are stronger than the object's self-gravity, and the object is pulled to pieces. The Roche limit differs for every planet, since their gravities vary, and it also differs for each orbiting object, since their densities differ. The Roche limit (R_L) can be calculated as follows:

$$R_L = 2.446\, R \left(\frac{\rho_{planet}}{\rho_{satellite}} \right)^{\frac{1}{3}},$$

where ρ_{planet} is the density of the planet, $\rho_{satellite}$ is the density of the object orbiting the planet, and R is the radius of the planet. Thus, moons that attempted to form within the Roche limit, or were thrown within the Roche limit by other forces, will be torn into rubble by the gravitational forces of the planet and form rings.

All the Uranian rings lie within the Roche limit, but Saturn's, Jupiter's, and Neptune's outer rings lie outside their Roche limits and orbit in the same regions as moons. The moons have important effects on the rings near which they orbit. First, if the moon and particles in the ring share an orbital resonance (when the ratio of the orbital times is an integer ratio, for example, the moon orbits once for every two times the particle orbits), they interact gravitationally in a strong way: The particle, moon, and planet line up regularly, exerting strong forces on the particle and warping its orbit. If the resonances are strong (low integer ratios), a gap can be created in the ring. In other cases the resonance results in a wavy ring.

Moons can also strongly affect ring particles that they orbit next to. Very thin rings that would otherwise be expected to widen with time can be kept thin by moons that orbit closely on either side. These are called shepherd moons, bumping any stray particles back into the rings or accreting them onto the moon's surface.

Intact moons outside the Roche limit may also shed material, forming the source of a ring. Small moons, with low gravity, may allow more material to escape than large moons do. Jupiter's small moons Adrastea, Metis, Amalthea, and Thebe are all thought to create their own rings. In the Saturnian system, by contrast, the large, 250-km radius moon Enceladus is thought to be the creator and sustainer of Saturn's E ring.

Once formed, the ring does not remain forever: Forces from radiation, meteoroid impacts, and drag from the outer parts of the planet's atmosphere (the exosphere) begin to erode the ring. It is estimated that even dense rings can only exist for a few hundreds of thousands or millions of years, and so even the gorgeous rings of Saturn are probably just a fleeting phenomenon in the age of the solar system. Faint rings may disappear within thousands of years, unless replenished by a moon.

A system of rings around a planet can be thought of as a miniature reenactment of the original solar nebula: The planet is the giant mass at the center of a rotating system, much as the early Sun was, and the rings are the material rotating around it. Material is taken from moons to make rings, and other material is swept up by moons, a sort of recycling between moons and rings. This may be the reason that the outer planets have rings and the inner planets do not; the outer planets have great inventories of moons to create and sweep up rings, while the inner planets do not have enough moons.

(continued from page 146)

thinnest rings are just 30 feet (10 m) thick. The seeming static calmness of the rings is simply a misleading consequence of viewing the rings from such an extreme distance. Though the large C, B, and A rings look continuous in most images, each ring is made up of dozens or hundreds of ringlets. Within each ring each particle collides with another once every few hours, probably occasionally accreting into short-lived fluffy clumps that are invisibly small from Earth.

Saturn's dark C and immense B rings are shown in the photo below. The major rings, those most visible and massive, are the C, B, and A rings, moving radially out from Saturn. The B and A rings are separated by Cassini's Division. Though the rings are bright, they are not massive: The total

Voyager 2 was 1.7 million miles (2.7 million km) from Saturn when it acquired these images. More than 60 bright and dark ringlets can be seen; the small squares are caused by the removal of reference marks during processing. (NASA/JPL/ Voyager)

mass of the C, B, and A rings is less than the total mass of the small moon Mimas, which has a radius of about 130 miles (209 km). The rings are bright because about 90 percent of the ring matter consists of pristine particles of water ice. The rings have some variation in albedo, implying that their composition also varies slightly. The main rings appear to consist of particles from about 0.4 inches (1 cm) to 16 feet (5 m) in diameter. The particle sizes for all the rings are inferred from their reflectivity and thermal emissions, since to date no single particle of the rings has been seen on its own.

In contrast to the bright and relatively massive C, B, and A rings, the E ring is thin and consists of particles in the micron range, a thousand times smaller than sand. The G ring particles are not so well sorted, and range from dust-sized up to kilometer-sized. The F ring is intermediate between the main rings and the E ring in terms of its density and particle size, but it is outstanding for its bizarre configurations. The F ring is narrow, multistranded, and inclined to the plane of the other rings. It contains kinks, clumps, and parts that appear braided; these structures continue to be the subject of study, and shocked scientists when first discovered by the Pioneer 11 mission. The F-ring kinks are caused by two "shepherd" moons, Prometheus and Pandora, each about 60 miles (100 km) in diameter. (For more on shepherd moons, see the sidebar "Why Are There Rings?" on page 144.)

The more that is learned about the rings and the better the images become, the more complex the rings' structures appear. Though Cassini's Division and Encke's Division appear empty from Earth, they are actually filled with very faint ringlets. Cassini's Division is maintained by the moon Mimas, and Encke's Division contains the orbit of the moon Pan. The larger rings contain waves and clumps and even smaller ringlets within their overall structures, in addition to the hundreds of tiny regular rings that spread evenly to make up the bulk of the major rings.

In the table on page 148, the rings are listed in order from closest to Saturn outward. Gaps and moons are also included in their proper order. Note that even though the Cassini and

SATURN'S RINGS

Ring, (moon), or division	Radial extent (miles [km])	Comments
cloud tops	37,470 (60,300)	
D ring	41,570 to 46,392 (66,900 to 74,658)	contains at least two narrow ringlets
C ring	46,392 to 57,152 (74,658 to 91,975)	contains at least four narrow ringlets, including the Titan and Maxwell ringlets
B ring	57,152 to 73,017 (91,975 to 117,507)	most opaque of rings
Cassini's Division	73,017 to 76,021 (117,507 to 122,340)	contains the Huygens ringlet
A ring	76,021 to 84,993 (122,340 to 136,780)	contains many small moonlets
Encke's Division	70,472 to 83,104 (113,410 to 133,740)	gap in A ring kept open by Pan
Keeler Gap	84,826 to 84,851 (136,510 to 136,550)	empty gap near edge of A ring; contains the moon Daphnis, discovered by the Cassini mission
(Atlas, Prometheus)	—	a ring 190 miles (300 km) wide is associated with Atlas
F ring	centered on 87,752 (141,220)	eccentric ringlet with kinks, clumps, and braid (Pandora, Epimetheus, Janus)
G ring	103,150 to 107,624 (166,000 to 173,200)	faint and isolated
E ring	111,800 to 298,000 (180,000 to 480,000)	broad faint dust ring including the orbits of Mimas, Methone, Anthe, Pallene, Enceladus, Tethys, Telesto, Calypso, Dione, Helene, and Polydeuces
Phoebe ring	3.7 to 7.4 (5.9 to 11.8 million)	immensely wide and thick low-density dust ring associated with Phoebe

Encke Divisions appear empty from Earth, they are actually filled with particles and even faint rings. They are simply much less dense areas than their surrounding rings.

Another surprising discovery from the *Voyager 1* mission was dark radial structures in the rings. These dark structures are about 5,000 miles (8,000 km) long and 1,200 miles (2,000 km) in width. They develop rapidly, over a matter of minutes, and disappear after several hours. While they last, the spokes travel with the rotation of the rings. Though they look dark in reflected light, they appear bright in scattered light, indicating that they are made of micrometer-sized particles. Electromagnetic fields may be responsible for these patterns of denser and thinner rings, since tiny particles are most easily moved by electromagnetic forces. There are a number of complex theories involving the planetary magnetic field to explain the formation of these spokes, but no one theory predominates at the moment.

Saturn's rings are very bright, which means the material in them is fresh and has not been covered by space-weathering products, indicating that the rings are geologically young. Uranus's and Neptune's, by comparison, are dark and red. Because the processes that destroy rings work so quickly, the rings cannot have existed for the age of the solar system and therefore cannot have formed at the same time as the other moons. It is thought that Saturn's rings are about 100 million years old, the result of an icy moon about 60 miles (100 km) in diameter being ripped apart by gravitational forces and scattered into rings, or perhaps they are the result of a comet capture and breakup. The ring material will eventually darken and redden, like those of the other planets. The rate of darkening by meteoroid impacts and plasma can be calculated, and this calculation results in the estimate that the rings are now about 100 million years old.

In October 2009, the *Spitzer Space Telescope* detected a new ring around Saturn—it has a very low density and is very far from the planet, which makes it unique among all planetary rings that have thus far been identified. The ring extends 3.7 to 7.4 million miles (5.9 to 11.8 million km) from the planet. In addition to its size, low density, and diameter, the ring is thick, as thick as 20 Saturn diameters. It is like a vast tube of dust around the planet. Anne Verbiscer, a research astronomer at the University of Virginia, reports that a billion Earths would fit in the volume of this ring. The sparse ring, however, contains only 10 to 20 particles per cubic quarter mile. Stranger still, the ring is orbiting in a retrograde sense, in the opposite direction from Saturn's other rings, and at an angle of about 27 degrees from the plane of the other rings. Saturn's moon Phoebe orbits with the new ring and is thought to be the source of the dust making up the new ring.

Rings consisting of tiny particles, such as the D, E, F, and G rings, will have the shortest lifetimes. Dust-sized particles are easily moved by drag, magnetic forces, and meteoroid impacts. These processes efficiently remove dust from the rings, and so dust must be replenished in the rings from a nearby source. Some scientists think that larger bodies, on the order of one kilometer in diameter, exist inside these dusty rings and continually shed dust into the rings to replenish them. Other scientists believe that by coincidence with the time of human space exploration, the rings are at the fleeting moment of their peak development. Dusty rings are only expected to exist for tens of millions of years, and denser rings for not significantly longer than 100 million years, and so ring systems are expected to form and dissipate many times over the age of the solar system.

SATURN'S MOONS

The standout of Saturn's population of moons is certainly giant Titan. Titan is a planet-sized moon, with a diameter about 40 percent the size of the Earth's. Though Jupiter's Ganymede beats Titan as the largest moon in the solar system, Titan is still larger than Mercury and Pluto, making it the ninth-largest body in the solar system. Titan has a denser atmosphere than either the Earth or Mars. Following Titan among Saturn's population of moons are four other moons with diameters about one-fifth of Titan's: Rhea, Iapetus, Dione, and Tethys, and from there the moons taper off in the end to tiny irregular satellites that are almost certainly captured asteroids.

In 1655 Christiaan Huygens found Titan, the first known moon of Saturn and the first known moon of any planet beyond Earth and Jupiter. The search for moons of the giant planets was on. Tethys, Dione, Rhea, and Iapetus were discovered by Cassini in 1684, 1684, 1672, and 1671 respectively, and then the rush of discovery stalled for a century. Mimas and Enceladus were both found by Herschel in 1789, and then another century passed before William C. Bond, an American watchmaker and amateur astronomer, and William Lassell, an English brewer and amateur astronomer, found Hyperion in 1848. The American astronomer William Pickering discovered Phoebe in 1898. In 1966 Richard Walker and Audouin Dollfus found Janus and probably Epimetheus. These two moons are co-orbital, but the clarification of their orbit and the discoveries of the next set of moons awaited the Voyager missions. In 1980 and 1981, thanks to both the *Voyager 1* and *2* missions and improved ground-based telescope instrumentation, Epimetheus, Helene, Telesto, Calypso, Atlas, Prometheus, Pandora, and Pan were all discovered. The remainder of the moons were discovered in 2000 or more recently, thanks to dedicated sky searches using new technology. As of 2009, there were 62 known moons of Saturn, but more will certainly be discovered.

A number of new Saturnian satellites were discovered in 2000 thanks to a special initiative by a team of astronomers. Brett Gladman, at the time a graduate student at Cornell University, and Cornell professors of astronomy Joseph Burns and Philip Nicholson conclusively identified four new moons of Saturn. Other members of the team included Jean-Marc Petit and Hans Scholl of the Observatoire de la Côte d'Azur, France; J. J. Kavelaars of McMaster University, Canada; and Matthew Holman and Brian Marsden of the Harvard-Smithsonian Center for Astrophysics.

The discovery of the four new moons was made using a technique implemented by Gladman while he was a student at Cornell. Gladman, who now works for the Centre National de la Recherche Scientifique in France, obtained his Ph.D. at Cornell. The technique, which also was used in the discovery of the five new Uranian moons, uses light-sensitive

(continues on page 154)

GIOVANNI CASSINI (1625–1712)

By no odd chance did NASA give the name Cassini-Huygens to its long, ambitious, important mission to Saturn. The spacecraft itself is named for Cassini, and the probe that will fly into the atmosphere of Saturn's moon Titan is named Huygens. In addition, a crater on Mars is named Cassini, and a dark region on Iapetus is called the Cassini Regio. Cassini was undeniably among the top astronomers in the history of humankind as a result of his astonishing array of discoveries. In brief, Cassini discovered:

* Jupiter's flattened poles and bulging equator, a result of its spin, in 1654
* Jupiter's bands and its Great Red Spot, in 1655 (the Red Spot may have been seen by Robert Hooke in 1664)
* Discrepancies in the occultation schedule of Jupiter's moons behind Jupiter itself that depended upon where Jupiter is in its orbit, in 1668; this led to the first calculation of the speed of light
* Saturn's moon Iapetus, in 1671
* Saturn's moon Rhea, in 1672
* The gap between Saturn's A and B rings, now named the Cassini Division, in 1675
* Saturn's equatorial belt, in 1676
* Saturn's moon Tethys, in 1684
* Saturn's moon Dione, in 1684
* Possibly the first human observation of a giant impact, on Jupiter, drawn from images seen at the Paris Observatory in 1690
* Differential rotation on Jupiter, around 1690

Cassini also calculated the rotational periods of Mars, Jupiter, and Saturn, and the orbits of Jupiter's moons.

Cassini was born Giovanni Domenico Cassini in Perinaldo, Genoa, Italy, in 1625. During his adulthood in Italy he sometimes used the name Gian Domenico Cassini, but when he moved to France he adopted the French form, Jean-Dominique Cassini. In college he was most interested in astrology, not astronomy; he studied astrology in depth but wrote that he was convinced there was no truth in astrological predictions. In 1644 the Marquis Cornelio Malvasia, however, invited Cassini to Bologna on the strength of his knowledge of astrology. Cassini was given a position at Malvasia's new Panzano Observatory, and he purchased the instruments for the observatory. In 1650 Cassini progressed to a chair in mathematics and astronomy at the University of Bologna, and his mature career was launched.

Cassini's knowledge of engineering and hydrology caused him to be invited by Pope Alexander VII to give advice on some disputes over the river Po. The pope so valued Cassini's advice that he was asked to take holy orders and work permanently for the pope, but Cassini declined, continuing to offer his services when asked. In 1668 Cassini was invited to Paris to work on building the new Paris Observatory, and in 1671 he was made its director. Cassini promptly became a French citizen and changed his name to its French version. Cassini married and settled in Paris; he had two sons, the younger of whom eventually succeeded him as director of the observatory.

Cassini's constant curiosity and hard work allowed him to conduct several large experiments, at one point sending scientists to measure longitudes all over the Earth using observations of eclipses of Jupiter's moons. As with any scientist, some of his ideas have proven with time to be incorrect: He proposed a form other than an ellipse for the orbit of planets, and his arc across France convinced him that the Earth is elongated at its poles, rather than flattened. In one of Cassini's most successful large experiments, his colleague Jean Richter traveled to French Guiana and made a measurement of the position of Mars simultaneously with Cassini in Paris. The two measurements allowed the men to measure the parallax to Mars, that is, the angle between the two viewers as they looked at Mars. This calculation, completed in 1672, gave a distance to Mars, the first true measurement of a distance in the solar system.

Cassini is viewed, in fact, as a second-rate theoretician but an observationalist of the first order. Making observations may seem the lesser achievement, but the exactness, care, and ability in engineering that makes great observational discoveries possible are specialties in their own right, and in these Cassini excelled.

As Cassini aged, his eyesight deteriorated, a devastating loss for an observational astronomer. By 1711 he was almost completely blind. Despite this cruel infirmity, Cassini remained gentle and positive. Cassini died in 1712, and his legacy in astronomy was carried on by several generations of his family.

(continued from page 151)
semiconductors, called charge-coupled devices, attached to telescopes to detect the distant points of light.

George Smith and Willard Boyle invented the charge-coupled device at Bell Laboratories in 1969, and once it was refined and put into mass production, it revolutionized cameras, fax machines, scanners, and, of course, telescopes. A charge-coupled device consists of many linked capacitors, electronic components that can store and transfer electrons. When a photon strikes the surface of the charge-coupled device, it can knock an electron off the atom in the surface it strikes. This electron is captured by the capacitors, one for each pixel of the image. While photographic film records a paltry 2 percent of the light that strikes it, charge-coupled devices can record as much as 70 percent of incident light. Their extreme efficiency means that far dimmer objects can be detected. This sensitivity made searching for small distant objects possible.

Several of these digital images, taken once every hour, are compared using computer software to pick out a moving point of light against the known star background of the sky. The first two candidates for newly discovered satellites of Saturn were spotted by Gladman using the European Southern Observatory's 2.2-meter telescope in Chile on August 7, 2000. Gladman and Kavelaars recovered (found again) the two objects September 23 and 24 at the Canada-France-Hawaii 3.5 meter telescope on Mauna Kea, Hawaii. They also found two new candidates. Additional confirming observations were made at other telescopes.

Their discoveries brought the known Saturnian moons to a total of 30, surpassing the 21 orbiting Uranus. Little is known about the four new moons except for their brightness. Estimates of their size (between six and 30 miles, or 10 and 50 km across) are based on assumptions of their reflectivity. Observed from Earth-bound observatories, the moons appear as faint dots of light moving around the planet. Between 1997 and 1999 the same team discovered a total of five new moons of Uranus. All five, like the newly discovered four outer moons of Saturn, are irregular satellites.

Twenty new moons were discovered in 2003 and 2004. The first, Saturn's 31st moon, S/2003 S1, was found February 5, 2003, by a University of Hawaii team led by Scott Sheppard and David Jewitt, along with Jan Kleyna of Cambridge University, using one of the ground-based telescopes on Mauna Kea. The two newest additions to the Saturnian family, provisionally named S/2004 S1 and S/2004 S2, were first seen by Sebastien Charnoz, a planetary dynamicist working with Andre Brahic, a Cassini imaging team member at the University of Paris in 2004 in images sent back to Earth by the Cassini mission. The satellites were the smallest moons of Saturn yet discovered, about two miles (3 km) and 2.5 miles (4 km) in diameter. They orbit within 131,000 miles (211,000 km) of the planet's center, meaning that they orbit between the orbits of Mimas and Enceladus, very close to the planet.

The remainder of the 20 new moons from those years and most of the 11 additional moons discovered in 2005, 2006, and 2007 orbit outside Iapetus. The moons inside Iapetus have nearly circular orbits that lie close to Saturn's equatorial plane. These inner moons with circular, prograde orbits are therefore likely to have formed at the same time Saturn did, while the outer moons with eccentric, inclined orbits are likely to be captured asteroids and are called irregular satellites. The inner satellites of Saturn, up to and including Iapetus, include 22 moons. Until the recent discoveries, Saturn was known to have only one irregular outer satellite, Phoebe, discovered by William Pickering more than 100 years ago. Phoebe travels in a retrograde orbit, that is, in the opposite direction to the spin of Saturn.

Saturn's most recently discovered irregular satellites bring the total of Saturn's irregular satellites to 38, while Jupiter has more than 50. Both planets almost certainly have more irregular satellites to be discovered. These small, dim, distant objects in eccentric and inclined orbits are difficult to see, and so discovery occurs slowly. Many of Jupiter's irregular satellites orbit in a retrograde direction, as do most of Saturn's. A retrograde orbit is virtually proof that the object was captured after planetary formation, since formation at the same time as the planet requires that the satellite orbit in the same direction that the planet spins.

Saturn's irregular satellites had appeared to fit into five groups according to their orbital inclinations. The groups appeared to orbit adjacent to each other and to share similar orbital inclinations, and so scientists thought these distinct groupings might indicate that the moons in each group came from a group of small bodies with similar speeds and directions prior to capture and may provide clues about where the captured satellites originated. Unfortunately, with the discovery of many new irregular satellites, the groupings have been disrupted and no longer appear to exist in the simple way they did with less data. Orbital inclinations skip up and down between values near 40 degrees and values of almost 180 degrees, and adjacent orbits also mix prograde with retrograde directions.

Tables of Jupiter's and Saturn's satellites change regularly as new satellites are discovered and as additional observations of existing satellites refine measurements of the orbit's semimajor axes, eccentricities, and inclinations, none of which are simple to calculate from astronomical observations. As the orbits become better known with more observations, perhaps clearer groups will emerge, but in the current understanding of the order of satellites there are no good groupings.

Saturn's major satellites are Mimas, Enceladus, Tethys, Dione, Rhea, Titan, Hyperion, and Iapetus, and of these, Tethys, Dione, Rhea, Titan, and Iapetus are by far the largest. The major satellites Mimas, Enceladus, Tethys, Dione, Rhea, and Iapetus, in addition to Epimetheus and Janus, are all in synchronous orbit with Saturn, meaning that their rotation period is the same as their revolution period, and they always show the same face to Saturn. The major satellite Titan is not in synchronous orbit. Hyperion's orbit is chaotic, and it and the satellites have not reached tidal locking with Saturn and so are not synchronous.

The innermost of Saturn's rings begins at about 41,000 miles (66,000 km), well inside the orbit of Saturn's innermost moon Pan, which orbits at about 83,000 miles (133,600 km). The Cassini mission, however, confirmed the existence of more than 150 tiny "propeller moonlets" within the ring inside Pan. These tiny bodies, none larger than 550 yards (500 m) in diameter, orbit in three gaps they have created in the inner A ring of Saturn.

All the moons through Helene, orbiting at about 235,000 miles (377,400 km), lie within the rings. Helene orbits Saturn with about the same orbital radius that the Moon has around the Earth. The outermost ring ends at 298,000 miles (480,000 km), inside the orbit of Rhea, at 328,000 miles (527,100 km). Between Rhea and Titan is a large empty region. Titan's orbital radius is more than twice Rhea's, at 759,000 miles (1,221,900 km). The outermost known moon is Fornjot, which has a highly eccentric orbit with a semimajor axis of 15,692,500 miles (25,108,000 km).

The table of Saturn's moons on page 164 lists all the known moons along with some of their physical and orbital characteristics. Following the table are sections for each of the moons, describing the origin of their names and what is known about them.

1. S/2009 S1

In 2009, a new moon around Saturn was found, the most recently discovered moon in the solar system. This tiny moonlet, less than two-tenths of a mile (0.3 km) in diameter, is embedded in the inner portion of Saturn's D ring, making it the moon closest to Saturn yet found. The moon is so small and has been observed for such a brief period of time that little is known about its physical or orbital properties.

2. Pan

Named after the Greek god of shepherds, flocks, and fertility, Pan is the innermost of Saturn's moons. It orbits in a tiny gap of its own making, the Encke division, in Saturn's A ring. Its gravitational influence creates small waves in the edges of the rings. An examination of these waves predicted the position and size of Pan, and Mark Showalter, a researcher at the Stanford Center for Radar Astronomy, found the moon in 1990 in nine-year-old *Voyager 1* photos. Pan seems to be icy body, but its small radius (about 6 miles or 10 km) prevents good measurements from Earth.

3. Daphnis

Discovered by the Cassini mission science team in 2005 and named for a poet and shepherd in Greek mythology, this moon makes gravitational ripples in the outer edge of the Keeler Gap in Saturn's rings.

4. Atlas

Atlas, named for the Greek Titan condemned by Zeus to hold the heavens on his shoulders in retribution for threatening the gods, is a shepherd body for Saturn's A ring. Richard Terrile of the Jet Propulsion Laboratory discovered Atlas in photos taken by *Voyager 1*. Almost nothing is known about this small moon, though the Cassini mission discovered a tiny ring probably created from dust shed by Atlas.

5. Prometheus

This moon is named for the Greek Titan Prometheus, the son of Iapetus, a Titan known as the father of mankind, and Clymene, the goddess of fame and infamy. Prometheus stole fire from the gods and gave it to humankind. For this awful crime Zeus chained Prometheus to a rock to be tortured by carrion birds for eternity (though he was later saved by Chiron). Prometheus was also the brother of Atlas and Epimetheus.

Prometheus is the innermost of the two shepherd moons that gives the F ring its kinks. Extremely elongated, with a long axis of about 92 miles (148 km) and a short axis of 42 miles (68 km), Prometheus is large enough that some additional information about it can be detected from Earth-based observations and from *Voyager* images. The moon seems to have ridges and valleys on its surface, along with several craters. Its surface appears smoother and therefore perhaps younger than its nearby companions Pandora, Janus, and Epimetheus. Like these neighbors, Prometheus's high albedo (0.6) and density lower than water ice imply that it may consist of porous or broken ice with cavities and fractures.

Stewart Collins and other members of the Voyager mission team found Prometheus in images from that mission taken in 1980. The *Voyager 1* data was extensive and precise enough to clearly define Prometheus's orbit, but when Prometheus was again measured by the spacecraft *Cassini* in 1995 and 1996, the moon was found to be 20 degrees behind in its orbit from its predicted position. Twenty degrees is an immense error, far beyond the uncertainties for calculations of Prometheus's orbit. In the intervening years between 1980 and 1995, Prometheus must have experienced some strong force, possibly either a collision with F ring material or interactions with a small companion moon that remains undetected.

6. Pandora

Pandora is named for the first human woman created by the Greek pantheon. Pandora was given to Epimetheus as his wife, along with the famous box of misfortunes that she could not, in her curiosity, prevent herself from opening.

Stewart Collins and his team also discovered Pandora from *Voyager 1* images. Pandora is the outer of the two "shepherd" moons that give Saturn's F ring its kinks. Pandora is of a similar size to Prometheus but is less elongated. Its long axis is 68 miles (110 km) and its short axis 38 miles (62 km). With what limited data there is on Pandora, a map can be made that implies the surface has no ridges or valleys, though it has several craters as large as 19 miles (30 km) in diameter. This moon has an albedo even higher than Prometheus's (0.9 compared to 0.6), though it has a similar density. Its composition may also be assumed to be ice with fracture or pore space, though in fact little is known about this moon.

7, 8. Epimetheus and Janus

Epimetheus was another Greek Titan, a brother of Prometheus and Atlas, and husband of Pandora. His task was to populate the Earth with animals. Janus, in contrast, is the two-headed Roman god of beginnings and doorways, and the guardian of the gateways to heaven. January, the month that looks back at the old year and on to the next, is also named for Janus.

Epimetheus is co-orbital (shares its orbit) with Janus. The American astronomer Richard Walker and the French astronomer Audouin Dollfus independently discovered Janus in 1966, and Walker probably detected Epimetheus at the same time. Twelve years later, in 1978, Stephen Larson and John

Fountain, astronomers at the University of Arizona, proved there were two moons in that orbit and so officially share the discovery of Epimetheus with Walker. The positions and sizes of the moons were completely clarified by the images taken by *Voyager 1* in 1980.

Epimetheus and Janus strictly share the same orbit, but at any given time one moon is orbiting about 30 miles (50 km) closer to Saturn than the other moon. One orbit lies at 94,123 miles (151,472 km) from Saturn, while the other is at 94,091 miles (151,422 km). Bodies in lower orbits travel faster than those in higher orbits. The moon in the lower orbit, then, will inevitably catch up to the other as they travel around Saturn. The radii of each of these moons is larger than their separation of 30 miles (50 km), but a collision is actually prevented by their mutual gravitational attraction. As the moon in the lower orbit approaches the other, its momentum is raised by the gravitational attraction between the moons. The moon in the lower orbit is raised into the higher orbit and slows down, while the moon in the higher orbit loses momentum, drops into the lower orbit, and speeds up. The moons begin moving apart again in their exchanged orbits without passing each other or even approaching very closely. Epimetheus and Janus make this approach and orbital switch about every four years. Until recently their arrangement was thought to be unique in the solar system, but some scientists now think that the Earth has co-orbital asteroids that switch orbits back and forth around the Earth rather than striking it.

Like the other nearby moons, Epimetheus and Janus both have high albedo, low density, and surfaces marked with large craters, ridges, and troughs. Their rough surfaces indicate that the moons have not been resurfaced for at least several billion years. Janus has fewer linear features on its surface than has Epimetheus, but they are probably similarly aged. Pandora's surface appears to be older than these moons' surfaces, but Prometheus's surface is younger. The origin of the moons' curious orbit is not known; the two may have originated as one larger satellite that was broken by a collision, but there is no specific data supporting or disproving this hypothesis. Their orbits also may be unstable over long periods of time.

9. Aegaeon

In ancient Greek religion, Aegaeon was one of three warlike giants, called the hundred-handed ones. It was discovered in 2009 from images taken by the Cassini mission. This moon orbits in a bright arc of the dusty G ring, the last of the dusty rings to have a moon found within it. It is so small that little is known about its physical or orbital characteristics.

10. Mimas

In ancient Greek religion Mimas was a giant. Like the Titans, the giants were all children of Gaia. After the Titans were defeated at war with the gods, Gaia incited the giants to fight the Olympians. Success could only be guaranteed the gods if they attracted humans to fight with them, and for that reason Hercules was brought into the battle and himself killed most of the giants. Mimas, however, was killed by molten metal thrown by the god Hephaestus and is now said to lie beneath Mount Mimas in Erythres, an ancient town now located in Turkey.

Mimas, pronounced "MY mass" or alternately "MEE mass," is the innermost of Saturn's major satellites. William Herschel, probably the most famous astronomer of the 18th century and the man who identified Uranus in 1781, discovered this bright, icy moon in 1789. It orbits at a distance of 116,250 miles (186,000 km), within Saturn's E ring. Mimas maintains Cassini's division in the rings and is responsible for the "corrugations" in the A ring. The Cassini division is in a 2:1 resonance with Mimas: Mimas orbits Saturn every 22 hours, while particles in and near the Cassini division orbit every 11 hours. Every two orbits any material that has drifted into the Cassini division is pulled toward Mimas by gravity and eventually cleared from the gap.

Mimas is heavily cratered, with craters about 20 miles (30 km) in diameter. The moon's outstanding feature is the crater Herschel, with a diameter of 81 miles (130 km), nearly one-third the size of the moon itself. This crater is six miles (10 km) deep and has a central peak four miles (six km) high. An impact this large would have been likely to destroy the moon; no clear explanation for its survival has been proposed. It is thought, though, that during the Late Heavy Bombardment at 4 billion years before present, many of the midsized satellites of Saturn were destroyed and reassembled. The surface antipodal to (opposite through the globe) Herschel is highly disrupted, indicating that the shockwaves created by the Herschel impact propagated through the whole moon.

Unlike the other moons orbiting near Saturn, Mimas is nearly spherical and not irregular. Mimas's radius is 121 miles (196 km). The moon is one of the smallest bodies in the solar system to have acquired a near-spherical shape. The intense heating and disruption of the impact that created Herschel may have been enough to allow the moon to reassemble into a spherical shape; the fact that there are no other craters are large as Herschel may indicate that the moon was completely disrupted by a previous, larger impact and then reassembled in time to receive the impact that created Herschel. Also, unlike the lighter interior moons, the density of Mimas is about 76 pounds per cubic foot (1,200 kg/m³), indicating that it consists largely of water ice, with some heavier components in addition.

11. Methone

Methone was discovered by the Cassini mission science team in 2004 and called initially S/2004 S1. This was the first new moon discovered by the Cassini mission and was followed quickly by the discovery of its neighbor Pallene. Methone's permanent name comes from one of seven daughters of the giant Alkyoneus in Greek mythology.

12. Anthe

Little is known about this tiny moon, discovered by the Cassini mission science team in 2007. Its name comes from Greek mythology and is another of the seven daughters of the giant Alkyoneus; Anthe is the 60th confirmed moon of Saturn and its newest at the time of writing.

13. Pallene

Soon after finding Methone in *Cassini* images, the team found this moon, provisionally called S/2004 S2. Pallene is another of the seven daughters of the giant Alkyoneus in Greek mythology.

14. Enceladus

Like Mimas, Enceladus is named for a giant in ancient Greek religion, a child of Gaia. While the giants were still compliant, the gods fought the Titans for ascendancy. After the gods prevailed, Gaia incited the giants to begin their own war, the Gigantomachy, with the gods. The goddess Athena killed Enceladus by crushing him with the island of Sicily, and his death throes caused the eruption of Mount Etna. The gods reached complete control over heaven and Earth with their eventual triumph over the giants.

Enceladus is a moon with a diameter of 306 miles (494 km) that orbits at 147,890 miles (238,000 km) from Saturn. William Herschel discovered Enceladus in 1789. Enceladus orbits at the center of Saturn's E ring, and detritus from Enceladus is thought to contribute to the ring. Enceladus has an orbit in resonance with Dione and Helene, which share an orbit exterior to Enceladus. Enceladus orbits twice for each time that Dione and Helen orbit once.

Voyager 1 took photos indicating at least four distinct types of terrain: cratered plains, and almost crater-free plains with extensive faults, grooves, and ridges, including Daryabar Fossa and Samarkand Sulci. (For more on these terms, see the sidebar "Fossa, Sulci, and Other Terms for Planetary Landforms.") The crater-free plains and Enceladus's very high albedo indicate that the moon was recently resurfaced, perhaps as recently as 100 million years ago.

Enceladus has what may be the highest albedo of any object in the solar system (~0.99, but also sometimes given as higher than 1 because of technical details of albedo calculations), indicating that its surface consists of fresh ice. Because Enceladus has a density of only 71 pounds per cubic foot (1,120 kg/m³), its resurfacing could not have been due to dense Earth-like silicate volcanism, but researchers had suggested its shiny surface was probably due to melting and spreading ices on its surface, and that the moon was therefore geologically active.

(continues on page 167)

SATURN'S MOONS

Moon	Orbital period (Earth days)	Radius (miles [km])	Year discovered	Orbital eccentricity	Orbital inclination
1. S/2009 S1	undeter-mined	0.2 (0.3)	2009	0.000	0.000
2. Pan	0.58	9 (15)	1981	0.000	0.001
3. Daphnis	0.59	4 (7)	2005	0.000	0
4. Atlas	0.60	14 by 12 by 6 (23 by 19 by 9)	1980	0.000	0.003
5. Prometheus	0.61	37 by 27 by 20 (60 by 44 by 32)	1980	0.002	0.008
6. Pandora	0.63	34 by 27 by 19 (55 by 44 by 31)	1980	0.004	0.05
7. Epimetheus	0.69	43 by 34 by 34 (69 by 55 by 55)	1966, 1980	0.021	0.335
8. Janus	0.69	60 by 59 by 48 (97 by 95 by 77)	1966, 1980	0.007	0.165
9. Aegaeon	0.808	0.15 (0.25)	2008	0.000	0.001
10. Mimas	0.94	130 by 122 by 119 (209 by 196 by 191)	1789	0.021	1.566
11. Methone S/2004 S1	1.01	~2 (~3)	2004	0.000	0.007
12. Anthe	1.04	~1 (~2)	2007	0.001	0.1
13. Pallene S/2004 S2	1.15	~2.5 (~4)	2004	0.000	0.181

Moon	Orbital period (Earth days)	Radius (miles [km])	Year discovered	Orbital eccentricity	Orbital inclination
14. Enceladus	1.37	159 by 153 by 152 (256 by 247 by 245)	1789	0.000	0.010
15. Telesto (leads Tethys in same orbit)	1.89	9 by 8 by 4.5 (15 by 12.5 by 7.5)	1980	0.001	1.158
16. Tethys	1.89	333 by 328 by 327 (536 by 528 by 526)	1684	0.000	0.168
17. Calypso (trails Tethys in same orbit)	1.89	9 by 5 by 5 (15 by 8 by 8)	1980	0.001	1.473
18. Dione	2.74	352 (564)	1684	0.000	0.002
19. Helene (leads Dione in same orbit)	2.74	11 by 10 by 9 (18 by 16 by 15)	1980	0.000	0.212
20. Polydeuces (follows Dione in same orbit)	2.74	~1 (~2)	2004	0.000	0.177
21. Rhea	4.52	475 (764)	1672	0.001	0.327
22. Titan	15.95	1,600 (2,575)	1655	0.029	1.634
23. Hyperion	21.28	115 by 87 by 70 (185 by 140 by 113)	1848	0.018	0.568
24. Iapetus	79.33	446 (718)	1671	0.028	7.57

(continues)

SATURN'S MOONS *(continued)*					
Moon	Orbital period (Earth days)	Radius (miles [km])	Year discovered	Orbital eccentricity	Orbital inclination
25. Kiviuq	449	~5 (~8)	2000	0.334	49.09
26. Ijiraq	451	~4 (~6)	2000	0.316	50.212
27. Phoebe	548	71 by 68 by 65 (115 by 110 by 105)	1898	0.164	173.0*
28. Paaliaq	686	~7 (~11)	2000	0.364	46.151
29. Skathi	728	~2 (~4)	2000	0.270	149.08*
30. Albiorix	783	~10 (~16)	2000	0.478	38.04
31. S/2007 S2	800	~2 (~3)	2007	?	176.68*
32. Bebhionn	835	~2 (~3)	2004	0.469	40.48
33. Erriapo	871	~3 (~5)	2000	0.474	38.11
34. Siarnaq	896	~12 (~20)	2000	0.295	45.6
35. Skoll	878	~2 (~3)	2006	0.464	155.62*
36. Tarvos	926	~4 (~7)	2000	0.531	34.68
37. Tarqeq	887	~2 (~3)	2007	0.160	49.9
38. Greip	921	~2 (~3)	2006	0.326	172.7*
39. Hyrrokkin	932	~2 (~4)	2006	0.333	153.27*
40. S/2004 S13	976	~2 (~3)	2004	?	167.4*
41. S/2004 S17	986	~1 (~2)	2004	?	166.88*
42. Mundilfari	956	~2 (~3)	2000	0.208	169.38*
43. Jarnsaxa	964	~2 (~3)	2006	0.216	162.86*
44. S/2006 S1	972	~2 (~3)	2006	?	154.23*
45. Narvi	1,004	~2 (~3)	2003	0.431	137.92*

Moon	Orbital period (Earth days)	Radius (miles [km])	Year discovered	Orbital eccentricity	Orbital inclination
46. Bergelmir	1,006	~2 (~3)	2004	0.142	157.38*
47. Suttungr	1,017	~2 (~3)	2000	0.114	175.8*
48. S/2004 S12	1,048	~1 (~2)	2004	?	164*
49. S/2004 S7	1,101	~2 (~3)	2004	?	165.6*
50. Hati	1,033	~2 (~3)	2004	0.372	163*
51. Bestla	1,084	~2 (~3)	2004	0.521	147.4*
52. Farbauti	1,086	~1 (~2)	2004	0.206	158.4*
53. Thrymr	1,094	~2 (~3)	2000	0.470	174.5*
54. S/2007 S3	1,100	~2 (~3)	2007	?	177.22*
55. Aegir	1,116	~2 (~3)	2004	0.252	167.24*
56. S/2006 S3	1,142	~2 (~3)	2006	?	150.8*
57. Kari	1,234	~2 (~3)	2006	0.478	148.38*
58. Fenrir	1,260	~1 (~2)	2004	0.136	162.83*
59. Surtur	1,298	~2 (~3)	2006	0.451	166.92*
60. Ymir	1,315	~6 (~9)	2004	0.335	172.1*
61. Loge	1,312	~2 (~3)	2006	0.187	166.54*
62. Fornjot	1,490	~2 (~3)	2004	0.206	167.89*

* indicates retrograde orbit

(continued from page 163)

In 2005, the Cassini mission made a tremendous discovery: The craft flew near Enceladus and took photographs that showed plumes of liquid erupting through the icy crust of Enceladus and flying as high as 50 miles (80 km) above the surface. This discovery created great interest in both the scientific community and with the public, because the plumes consist mainly of water, and where there is water, there may be life. Enceladus became a more central target for research in the Cassini mission, and indeed many of the mission's goals for its operation beyond its initial planned time span concern Enceladus. In 2008, the *Cassini* flew just 30 miles (48 km) above Enceladus and took the highest-resolution photos yet obtained. These photos show that the plumes originate in cone-shaped pits up to 1,000 feet (300 m) deep. Mounded deposits near the cones are thought by mission scientists to be piles of snow left by the geysers as they freeze.

The plumes originate at warm plots near Enceladus's south pole. Enceladus's average surface temperature is about -324°F (-200°C), and parts of the planet can be as cold as -400°F (-240°C). In the warm spots, the temperature rises to an average of -306°F (-188°C), and in places as high as -207°F (-133°C), over 115°F (65°C) warmer than the moon's average temperature. These are still very low temperatures for liquid to exist, particularly water, which freezes at 32°F (0°C), and some scientists have suggested that this melting is made possible by the combination of ammonia and water. Ammonia is a volatile chemical, that is, it easily forms a gas. The temperature at which ammonia forms a liquid from an ice in the presence of water is also very low, just -142°F (-97°C).

This lowest temperature at which water and ammonia together can remain as a liquid is still warmer than the warmest temperature measured on Enceladus, but the warm temperatures on Enceladus were measured at the surface—could the moon be warmer inside, where the liquid is originating? This is the scientists' hypothesis.

Perhaps the largest remaining mystery about Enceladus is the origin of the heat that melts the ices into liquid. To create the warm region near the south pole where the geysers originate, some heating mechanism must provide about 6 GW of energy to Enceladus. The two primary hypotheses for heating this tiny moon are radioactive decay and tidal heating. Most of the significant heat-producing radioactive elements chemically bond with silicate rocks and not with watery ices and so are concentrated in the inner solar system with Mercury, Venus,

Earth, and Mars. There are no models of formation of these icy moons that indicate they should have any significant radioactive heating, and no remote sensing techniques have provided any data showing significant radioactive element concentrations. The best estimate for Enceladus's rate of radiogenic heating is 0.32 GW, about 18 times too small to account for the geysers.

Tidal heating, therefore, has been the best hypothesis for heating Enceladus. Jennifer Meyer and Jack Wisdom from MIT have done a careful analysis of all sources of tidal heating for Enceladus (that is, not just from Saturn itself, but also from orbital interactions with other moons, notably from the 2:1 resonance with Dione). They have found that the maximum tidal heating rate for Enceladus is only 1.1 GW, only one-sixth of the necessary heat.

Thus the fascinating water geysers on Enceladus remain a mystery—there is no reasonable hypothesis for the source of heat that powers them. Enceladus is not alone in this problem. Jupiter's moon Io, which is continually volcanically active, also required more power than can be theoretically provided by tidal heating. Though it remains unexplained, the excess heat in these moons makes them particularly good candidates in the search for extraterrestrial life.

16. Tethys

Tethys (pronounced "TEE this") is named for a Titan in ancient Greek mythology who was goddess of the fertile oceans. Tethys married her brother Oceanus and bore 3,000 children. The daughters of this family were the Oceanids, the festive sea nymphs. Tethys is aptly named because it is probably made almost entirely of water ice, since its density is 71 pounds per cubic foot (1,120 kg/m³), slightly higher than water ice. Its water-dominated composition and surface characteristics make it similar to the moons Dione and Rhea. Tethys has a surface temperature of -305°F (-187°C), slightly warmer than Enceladus, probably because Tethys has slightly lower albedo (~0.9) that allows more solar heating of its surface.

In 1684 Giovanni Cassini discovered Tethys. Its diameter is 656 miles (1,056 km), and it orbits an average of 183,309 miles

(continues on page 172)

FOSSA, SULCI, AND OTHER TERMS FOR PLANETARY LANDFORMS

On Earth the names for geological features often connote how they were formed and what they mean in terms of surface and planetary evolution. A caldera, for example, is a round depression formed by volcanic activity and generally encompassing volcanic vents. Though a round depression on another planet may remind a planetary geologist of a terrestrial caldera, it would be misleading to call that feature a caldera until its volcanic nature was proven. Images of other planets are not always clear and seldom include topography, so at times the details of the shape in question cannot be determined, making their definition even harder.

To avoid assigning causes to the shapes of landforms on other planets, scientists have resorted to creating a new series of names largely based on Latin, many of which are listed in the following table, that are used to describe planetary features. Some are used mainly on a single planet with unusual features, and others can be found throughout the solar system. Chaos terrain, for example, can be found on Mars, Mercury, and Jupiter's moon Europa. The Moon has a number of names for its exclusive use, including lacus, palus, rille, oceanus, and mare. New names for planetary objects must be submitted to and approved by the International Astronomical Union's (IAU) Working Group for Planetary System Nomenclature.

NOMENCLATURE FOR PLANETARY FEATURES

Feature	Description
astrum, astra	radial-patterned features on Venus
catena, catenae	chains of craters
chaos	distinctive area of broken terrain
chasma, chasmata	a deep, elongated, steep-sided valley or gorge
colles	small hills or knobs
corona, coronae	oval-shaped feature
crater, craters	a circular depression not necessarily created by impact
dorsum, dorsa	ridge
facula, faculae	bright spot
fluctus	flow terrain
fossa, fossae	narrow, shallow, linear depression

Feature	Description
labes	landslide
labyrinthus, labyrinthi	complex of intersecting valleys
lacus	small plain on the Moon; name means "lake"
lenticula, lenticulae	small dark spots on Europa (Latin for freckles); may be domes or pits
linea, lineae	a dark or bright elongate marking, may be curved or straight
macula, maculae	dark spot, may be irregular
mare, maria	large circular plain on the Moon; name means "sea"
mensa, mensae	a flat-topped hill with cliff-like edges
mons, montes	mountain
oceanus	a very large dark plain on the Moon; name means "ocean"
palus, paludes	small plain on the Moon; name means "swamp"
patera, paterae	an irregular crater
planitia, planitiae	low plain
planum, plana	plateau or high plain
reticulum, reticula	reticular (netlike) pattern on Venus
rille	narrow valley
rima, rimae	fissure on the Moon
rupes	scarp
sinus	small rounded plain; name means "bay"
sulcus, sulci	subparallel furrows and ridges
terra, terrae	extensive land mass
tessera, tesserae	tile-like, polygonal terrain
tholus, tholi	small dome-shaped mountain or hill
undae	dunes
vallis, valles	valley
vastitas, vastitates	extensive plain

(continues)

(continued)

The IAU has designated categories of names from which to choose for each planetary body, and in some cases, for each type of feature on a given planetary body. On Mercury, craters are named for famous deceased artists of various stripes, while rupes are named for scientific expeditions. On Venus, craters larger than 12.4 miles (20 km) are named for famous women, and those smaller than 12.4 miles (20 km) are given common female first names. Colles are named for sea goddesses, dorsa are named for sky goddesses, fossae are named for goddesses of war, and fluctus are named for miscellaneous goddesses.

The gas giant planets do not have features permanent enough to merit a nomenclature of features, but some of their solid moons do. Io's features are named after characters from Dante's *Inferno*. Europa's features are named after characters from Celtic myth. Guidelines can become even more explicit: features on the moon Mimas are named after people and places from Malory's *Le Morte d'Arthur* legends, Baines translation. A number of asteroids also have naming guidelines. Features on 253 Mathilde, for example, are named after the coal fields and basins of Earth.

(continued from page 169)

(295,000 km) from Saturn. Tethys is the largest of Saturn's inner satellites. Tethys is co-orbital with Calypso and Telesto, which orbit in Tethys's Lagrange points, 60 degrees ahead and 60 degrees behind Tethys itself (for more on Lagrange points, see the sidebar "Stable Orbits for Small Bodies among Much Larger Planets" on page 176).

The northern hemisphere of Tethys is heavily cratered and bears in particular the huge impact structure Odysseus, with a diameter of 273 miles (440 km), about 40 percent of the diameter of the moon itself. Unlike craters on the Moon and other brittle planets, Odysseus has no deep depression, and in fact it has risen to conform to the curve of the Tethys globe. Odysseus must have begun as a depression with a peak ring and circling ridges, but apparently Tethys can flow sufficiently to remove these features over time. Tethys's ability to flow also helps explain why the moon was not completely destroyed by an impact this size, as it would have been had it been brittle. At the time of the impact that caused Odysseus, Tethys would have had to be largely malleable to withstand the impact's disruptive forces.

Tethys shows two distinct types of terrain: bright, densely cratered regions, such as that near its north pole, and relatively dark, lightly cratered planes that extend in a broad belt across the satellite. The densely cratered terrain is believed to be part of the ancient crust of the satellite, and the lightly cratered planes are thought to have been formed later by internal processes that resurfaced older cratered regions.

The opposite hemisphere is less cratered, and it is dominated by a plain cut by a huge valley system with multiple faults called the Ithaca Chasma. (For more on terms such as *chasma,* see the sidebar "Fossa, Sulci, and Other Terms for Planetary Landforms" on page 170.) This valley system is about 60 miles (100 km) wide at its widest, two to three miles (3 to 5 km) deep, and around 1,200 miles (2,000 km) long, reaching two-thirds of the way around the moon. With its exceptionally cold surface temperature, Tethys would never have had flowing liquid to carve a chasm, and so Ithaca Chasma must have been formed by tectonic forces, perhaps linked with the impact that caused Odysseus.

15, 17. Telesto and Calypso

The moon Telesto is named for one of the Greek Oceanid nymphs of Greek mythology, a daughter of Tethys and Oceanus. She is the symbol of success. Calypso, a nymph daughter of Titan Atlas, is the goddess of silence. She has a love affair with the shipwrecked Odysseus, and after his rescue, dies of grief.

Bradford Smith, an astronomer from the University of Arizona, and his colleagues Harold Reitsema, Stephen Larson, John Fountain, Dan Pascu, P. Kenneth Seidelmann, William Baum, and Douglas Currie discovered these two tiny moons from ground-based observations made in 1980. Smith and the rest of the team were able to see the small objects because they were using the prototype instruments for the *Hubble Space Telescope.* The long axes of each of these nonspherical moons is only about 18 miles (29 km) long, making them among the smallest moons in the solar system known until the recent discoveries of many tiny irregular satellites orbiting at great distances from both Saturn and Jupiter.

Telesto and Calypso are co-orbital satellites of the moon Tethys. Unlike Epimetheus and Janus, all three of these moons orbit at the same rate and therefore with the same period. This is possible because Tethys is so much more massive than Telesto and Calypso. Tethys creates two stable places in its orbit, called Lagrange points, where smaller bodies can orbit indefinitely without disturbance. There are other Lagrange points associated with a larger body (see the sidebar "Stable Orbits for Small Bodies among Much Larger Bodies" on page 176), but these two are 60 degrees ahead (Telesto) and 60 degrees behind (Calypso) Tethys in its orbit. The first examples found of solar system bodies in these theoretically stable orbits were the populations of asteroids that orbit ahead of and behind Jupiter, called the Trojan asteroids. In reference to these, Telesto and Calypso are sometimes called the Tethyan Trojans.

18. Dione

In ancient Greek mythology, Dione was an early goddess with a poorly documented role and history. In some legends she is an Oceanid nymph daughter of Oceanus and Tethys, but in others she is a Titan daughter of Gaia. She may also have been the mother of Aphrodite as well as Niobe and Pelops. Dione is also the feminine form of Zeus.

In 1684 Giovanni Cassini discovered Dione, a moon with a diameter of 696 miles (1,120 km). It orbits at 234,260 miles (377,000 km) from Saturn and has a density of 95 pounds per cubic foot (1,490 kg/m³), the densest of all the satellites except Titan. This density indicates a mixed composition of ices and rocks. Dione shares its orbit with the tiny moon Helene, which orbits in the Lagrange point 60 degrees ahead of Dione. Dione and Helene are in a 1:2 resonance with the moon Enceladus (Dione and Helene orbit once for every two times that Enceladus orbits).

Unlike Rhea and Mimas, Dione has a wide range of crater sizes. Dione has heavily cratered plains with many craters larger than 60 miles (100 km) in diameter, as well as plains with fewer craters and areas with almost no cratering at all; these features can be seen in the *Voyager* image above. Dione's craters have relaxed to relative flatness, as has Mimas's giant crater Odysseus. Dione's largest crater is Amata, with a diameter of 144 miles (231 km). Bright rays and wispy material around Amata are clearly visible in this image. Some of these deposits may be ejecta rays from Amata, but others are curving and not radial to the crater. These rays are some of the brightest features on any of Saturn's moons, and they are thought to be frost deposits. The curving rays may be frost deposits associated with crustal fractures, perhaps still caused by the impact but not created by ejecta.

Saturn's moon Dione displays bright ejecta rays from its crater Amata. (NASA/JPL/ Voyager)

Dione is tidally locked to Saturn, meaning that the same face is oriented forward in its orbit at all times. Contrary to expectations, Dione's leading hemisphere is the least cratered, while its trailing hemisphere has some of the most heavily cratered regions. Many of Dione's craters, however, would be large enough to break the moon's tidal lock with Saturn and cause it to turn and eventually relock in a new orientation. Dione's tidal lock may have been changed in this way several times in its history, ending with its current position with the most cratered region to the trailing side. Alternatively, Dione has remained in its current locked position for the age of the solar system, but its leading edge was preferentially resurfaced by *cryovolcanism*. Dione has apparently been in this orientation for billions of years based on the coating of dust on its leading surface. The dusty leading surface has a higher albedo than has the trailing face. Rhea's plains are also crossed with troughs, fissures, and features like the Moon's sinuous rilles, which on the Moon are thought to be collapsed lava tubes.

The longest fissure is Palatine Chasma, at 245 miles (394 km) long and up to five miles (8 km) wide.

19. Helene

In ancient Greek mythology, Helen was the daughter of Leda and Zeus, and the sister to Castor, Pollux, and Clytemnestra. Helen was the most beautiful of women and her abduction by Paris to Troy initiated the Trojan war. The moon Helene shares

STABLE ORBITS FOR SMALL BODIES AMONG MUCH LARGER PLANETS

The gravitational fields of planets as they orbit the Sun are strong forces on small bodies like asteroids, and likewise the gravitational fields of large moons as they orbit their planets are strong forces on the smaller moons. Finding a stable orbit where the gravitational forces of larger moons do not pull the smaller moons into a collision, throw them into the planet, or fling them out of orbit entirely is surprisingly difficult. The two major ways that smaller moons can orbit near larger moons, and that asteroids can safely orbit near planets, are through resonance of orbits, and by orbiting at Lagrange points.

RESONANCE

Orbital periods can differ from one another by an integer multiple. For example, the period of one orbit is two years, and the other six, differing by a multiple of three; this means that every three years the two bodies will have a close encounter. Even asteroidal orbits that differ by an integral multiple from the orbit of the large planet are usually cleared of asteroids by the gravitational field of the large planet, but some special integral multiples of orbits actually stabilize the orbits of the asteroids, maintaining them in place indefinitely. These stabilizing multiples are called resonances, and the unstable multiples eventually are depopulated of asteroids, creating gaps. The Jupiter system has good examples of both resonances and gaps. One group of Jovian asteroids, the Trojans, are at 1:1 (meaning that their orbit and Jupiter's are the same). Another group, the Hildas, are at 3:2 (meaning they orbit the Sun three times for every two Jupiter orbits), and the asteroid Thule orbits at 4:3. There are gaps at 1:2, 2:3, 1:3, 3:1, 5:2, 7:3, 2:1, and 5:3. Gaps in the asteroid belt which are cleared out by gravitational interactions with Jupiter have the special name of Kirkwood gaps because they were first observed in 1886 by Daniel Kirkwood, a professor of mathematics at Indiana University in the second half of the 19th century.

Dione's orbit, staying 60 degrees ahead of the larger moon in one of its stable Lagrange points. Helene is an irregular satellite with a long axis of about 22 miles (35 km). Very little is known about Helene.

20. Polydeuces

This tiny moon was discovered by the Cassini mission science team in 2004 and called S/2004 S5. This tiny moon shares its

LAGRANGE POINTS

Joseph-Louis Lagrange, a famous French mathematician who lived in the late 18th and early 19th centuries, calculated that there are five positions in an orbiting system of two huge bodies in which a third small body, or collection of small bodies, can exist without being thrown out of orbit by gravitational forces. More precisely, the Lagrange points mark positions where the gravitational pull of the two large bodies precisely equals the centripetal force required to rotate with them. In the solar system the two large bodies are the Sun and a planet, and the smaller body or group of bodies, asteroids.

Of the five Lagrange points, three are unstable and two are stable. The unstable Lagrange points, L1, L2 and L3, lie along the line connecting the two large masses. The stable Lagrange points, L4 and L5, lie within the orbit of the planet, sixty degrees ahead and sixty degrees behind the planet itself.

The L4 and L5 points are stable orbits so long as the mass ratio between the two large masses exceeds 24.96. This is the case for the Jupiter-Sun, Earth-Sun, Earth-Moon, and Tethys-Saturn systems, and for many other pairs of bodies in the solar system. Objects found orbiting at the L4 and L5 points are often called Trojans, after the three large asteroids Agamemnon, Achilles, and Hektor that orbit in the L4 and L5 points of the Jupiter-Sun system. (According to Homer, Hektor was the Trojan champion slain by Achilles during King Agamemnon's siege of Troy).

There are hundreds of Trojan-type asteroids in the solar system. Most orbit with Jupiter, but others orbit with Mars. The first Trojan-type asteroid for Mars was discovered in 1990 and named 5261 Eureka. In addition, several of Saturn's moons have Trojan companions. No large asteroids have been found at the Trojan points of the Earth-Moon or Earth-Sun systems. However, in 1956 the Polish astronomer Kordylewski discovered dense concentrations of dust at the Trojan points of the Earth-Moon system.

orbit with Dione, trailing Dione in its Lagrangian point, similarly to the way the moon Calypso trails Tethys in its same orbit, and paired with Helene, which leads Dione in its other Lagrangian point. Polydeuces, however, has an orbit that is perturbed by other gravitational pulls, and it wandered from 26.1 degrees behind Dione to as far as 31.4 degrees behind Dione.

21. Rhea

In ancient Greek mythology, Rhea was a Titan, both the wife and sister of Kronos (Saturn to the Romans). Her children were important figures: With Kronos, she had Zeus (Jupiter to the Romans), Poseidon (Neptune to the Romans), Pluto, Hestia, Hera, and Demeter.

Rhea is the second largest of Saturn's moons after Titan, with a diameter of 949 miles (1,528 km), but it still contains less than 2 percent of Titan's mass. Giovanni Cassini discovered the moon in 1672. Rhea orbits Saturn at a distance of 327,470 miles (527,000 km). Rhea's density (84 lb/ft^3, or 1,330 kg/m^3) indicates that the moon probably has a rocky core making up about one-third of the moon's mass. As a result of its albedo of 0.7, Rhea's surface temperature is -281°F (-174°C) in direct sunlight and between -328°F and -364°F (-200°C and -220°C) in the shade.

Voyager 1 took good pictures of its surface, which show that unlike the other moons of Saturn, Rhea has a surface relatively uniformly covered with craters. The majority of the craters are around 12 miles (20 km) in diameter, with the exception of the crater named Izanagi, which has a diameter of 155 miles (250 km). Rhea is tidally locked to Saturn, and its leading face is more heavily cratered than its trailing face (this is the expected distribution but opposite to what Dione displays). Like Dione, Rhea has a leading face brightened by a coating of dust and bright wispy features on its trailing side.

Because Rhea has no orbital resonance with other moons, it has cooled and is geologically inactive, resulting in a surface with few scarps, faults, or resurfacing. Though Rhea is thought to have an icy surface, its craters are still sharp and deep, unlike the relaxed craters in the flowing ice of Enceladus, Jupiter's moons.

22. Titan

In ancient Greek religion, the six female and six male Titans, the children of Uranus and Gaia, were Kronos (Saturn to the Romans), Iapetus, Hyperion, Oceanus, Coeus, Creus, Theia, Rhea, Mnemosyne, Phoebe, Tethys, and Themis. The name Titan is sometimes used for their descendants, including Prometheus, Atlas, Hecate, Selene, and in particular Helios. The Titans, led by Kronos, deposed Uranus and ruled the universe in early times. The Olympians, led by Zeus, overthrew the Titans in the battle called the Titanomachy.

Christiaan Huygens discovered the moon Titan in 1655. Titan was the only Saturnian moon found before the space age. In 1908, the Catalan astronomer José Comas Solà saw darker light at the edge of Titan and suggested it had an atmosphere, an amazing early observation from Earth. The moon has been photographed by several space missions and finally visited by a probe.

Voyager 1 passed 2,730 miles (4,394 km) from Titan in 1980, taking good images, and *Voyager 2* passed by about 100 times farther away about nine months later. Finally *Cassini* made a real visit to Titan with a series of 44 flybys starting in October 2004 and continuing for five more flybys in an extended mission after the primary mission ended on June 30, 2008. *Cassini* also dropped the *Huygens* probe to the surface of Titan; the probe took atmospheric measurements as it fell and photographed the surface.

Titan, like the planet Venus, is covered with clouds to the point that its surface can never be seen in visible light. The moon looks like a reddish-orange ball of clouds. These clouds are different north and south of Titan's equator: The northern clouds are brighter, by 25 percent in the blue light wavelengths but only a few percent in red and ultraviolet. It is thought that the hemispherical difference may be due to atmospheric circulation patterns. There is also a dark blue or purple hood over Titan's north pole from 70 degrees to 90 degrees north latitude, but only in its winter months.

Titan is the only one of Saturn's moons that can be seen as a disk from Earth. The other Saturnian moons appear as points; they are 100 to 100 million times less massive than Titan. Titan therefore contains more than 90 percent of the mass orbiting Saturn, including the rings. Titan's orbit is highly eccentric, with an eccentricity of 0.028. This eccentricity is too large to have been caused by tidal disruptions by other bodies, and so it is thought that Titan may have recently sustained a large meteorite or comet impact that disturbed its orbit. Unlike almost all other moons, Titan has its own magnetic field that is less than 10^{-7} Tesla at the surface.

With a diameter of 3,200 plus or minus two miles (5,150 $\pm$ 4 km), Titan is smaller than Jupiter's Ganymede but larger than Mercury (though less than half its mass). Titan is larger than the Moon and Pluto but smaller than Mars. Titan is therefore the second largest moon in the solar system after Ganymede; Titan is a planet-sized moon. Titan's surface gravity is 4.4 feet per second per second (1.35 m/sec^2), which is only slightly less than the Moon's (5.2 ft/sec^2 or 1.6 m/sec^2). Titan's size and mass indicate that it is about half rock and half ice.

Nearly 10 times as far from the Sun as the Earth is, Titan has a cloud temperature of -290°F (-179°C). The "greenhouse" of Titan's thick atmosphere warms its surface (the atmosphere traps the heat of sunlight, rather than letting it radiate back out to space), but Titan is also cooled by organic smog created by ultraviolet light striking the atmosphere (the smog bounces sunlight back out, preventing it from heating the moon). These two characteristics of its atmosphere are shared by Earth, which also has both greenhouse warming gases and smog and aerosols that cool the surface.

Titan is unique among moons in having a substantial atmosphere. Titan is one of only four solar system bodies that have a nitrogen-rich atmosphere: the first is the Earth, the second is Neptune's moon Triton, and Pluto also has a sporadic, thin, nitrogen-based atmosphere. The second largest constituent of Titan's atmosphere is methane, at a few percent of the atmosphere. Titan's atmosphere also contains carbon dioxide, a few 10s of parts per billion water, and some simple hydrocarbons in addition to methane: acetylene, ethylene, propane, and ethane. There are a few hundreds of parts per billion of water in its upper atmosphere (most water is thought to be on the surface, in the form of ice). Methane is expected to form clouds and rains just as water vapor does on Earth. Titan's nitrogen-based atmosphere is very similar to the early Earth's, before oxygen began to be produced by living organisms.

Because of its very thick atmosphere, Titan was the last object in the solar system with a surface that is largely unknown. The *Hubble Space Telescope* can "see" through the atmosphere by using specific wavelengths in the near-infrared that are only weakly absorbed by the methane in the atmosphere and so built up maps of Titan's surface showing light and dark areas (more on this below). Not until the *Huygens* probe landed on Titan in 2005 were images of the surface available.

Titan's atmosphere is thicker than those of Earth, Mars, and Mercury. The pressure at the surface of Titan is 1.45 bars, actually higher than Earth's 1.0 bars. Titan's atmosphere is much deeper than the Earth's, as well: its depth is similar to the radius of Titan itself, while the Earth's is more like the shell on an egg, largely because the Earth's much stronger gravity holds the atmosphere closer to its surface. In both Titan's atmosphere and the Earth's, the temperature falls with altitude up to a certain height, beyond which the temperature rapidly rises. This point, the temperature inversion, is called the tropopause, and the atmospheric layer below it is called the troposphere.

On the Earth, the height of the tropopause varies from about four to 10 miles (six to 16 km), while on Titan, it is at about 25 miles (40 km). The lowest temperature at the tropopause on Titan is about -330°F (-200°C), while the surface temperature is about -290°F (-180°C). At the cool tropopause most of the constituents of Titan's atmosphere including methane, ethane, and nitrogen may condense from a gas phase to droplets, just as water does in the Earth's atmosphere to form clouds. The surface temperature of Titan is low enough that water cannot sublimate and interact with Titan's atmosphere but instead stays frozen on the surface like rocks on Earth. On Titan, its condensates create a haze and possibly clouds, and this thick atmosphere is why Titan's surface was the last surface in the solar system that remained almost entirely unknown.

The middle atmosphere above the tropopause is called the stratosphere. In the stratosphere, the temperature falls at first and then remains steady. On Titan, the temperature of the stratosphere is about -148°F (-100°C), and the stratosphere is about 310 miles (500 km) thick. Having a hotter surface than tropopause is the result of the greenhouse effect, in which a layer of a particular gas in the stratosphere absorbs solar energy and radiates it downward to heat the troposphere, while insulating the troposphere from losing its heat again. On the Earth, the heating effect is created by the ozone layer in the stratosphere, while on Titan the deep layer of aerosols provides the heating effect.

Titan's temperature profile and radius were measured by *Voyager 1. Voyager 1* conducted an experiment called radio-occultation, to measure Titan's radius and surface temperature and pressure. In this experiment, *Voyager* measured radio frequency emissions from Titan as it passed behind Saturn. By measuring the time it took for Titan to pass behind, *Voyager* could determine Titan's diameter.

The origin of Titan's atmosphere is a bit of a mystery: Both Jupiter and Saturn are thought to have produced so much heat during their formations that their heat radiation blew gases away from their moons. Titan, then, may have acquired an atmosphere later, perhaps by outgassing, release of gases from its solid interior. The cold outer solar system temperatures then allowed Titan to keep its atmosphere; heat would have allowed the atmospheric gases to escape Titan's gravity. Why, though, so much atmosphere? The density of Titan's atmosphere is unexplained.

One hypothesis suggests that Titan sustained some cometary impacts, but measurements of the ratio of deuterium (hydrogen with an extra neutron) to hydrogen on the comets Halley, Hyakutake, and Hale-Bopp indicate that their ratios are four or more times higher than the ratio of these elements in methane on Titan. Very little of Titan's atmosphere, therefore, can have been contributed by comets. Another school of thought holds that Titan was able to hold on to an early nitrogen atmosphere even as it formed. According to calculations, massive Ganymede orbiting around Jupiter just that much closer to the Sun would not be able to hold on to its atmosphere because of solar heating.

Titan revolves around Saturn in 16 days at a distance of 1.2 million km, three times the Earth–Moon distance (this orbital distance is equivalent to 20.4 Saturnian radii). Titan's four seasons progress during Saturn's 29-year orbit around the Sun. In its cold season, Titan may even experience hydrocarbon snow. Since Titan has no magnetic field of its own, Saturn's magnetic field affects it strongly. Ions and electrons moving in Saturn's magnetic field bombard Titan's upper atmosphere and ionize and pull away atmospheric gases, mainly hydrogen, because of its small mass.

The moon long has been thought to have liquid hydrocarbon oceans, which were of great interest because until now Earth was the only planet in the solar system known to have liquid oceans. Atmospheric smog prevented *Voyager 1* from obtaining good surface photos of Titan during its close flyby in 1980, but recent crude maps made by the *Hubble Space Telescope* and ground-based telescopes showed large dark regions and highly reflective regions that may indicate ethane and methane oceans and ice continents.

A number of scientific papers were written hypothesizing about oceans of methane or ethane, but no clear evidence of their existence had been found. Other scientists studied the methane-ethane-water ice system and concluded that there is not enough methane in the atmosphere to coexist with a pure methane ocean, because more evaporation would be expected from an ocean of methane, leading to more methane in the atmosphere. Computer models of the chemicals at the temperatures expected on Titan predict that several hundred meters of liquid ethane should have accumulated on the planet's surface.

Based on all these lines of evidence scientists expected *Huygens* to find liquid oceans. When the probe landed, however, it photographed a surface covered with networks of channels apparently created by flowing liquids and dark low-lying areas that were probably seas of some kind. At the time of landing the channels and lowlands were dry.

Emily Schaller, Mike Brown, and H. Roe, scientists at the California Institute of Technology, with collaborators at the Gemini and Keck Observatories, have developed a technique to detect when large methane clouds form on Titan. The moon has distinct outbursts of large, bright methane clouds. Clouds appear in bursts at Titan's south pole and in streaks elsewhere in the southern atmosphere. The team from CalTech hypothesized that during Titan's summer methane is heated to the point that it evaporates and forms brilliant clouds and that in the fall and winter methane rains out of the clouds to refill the channels and lowlands on the surface of the moon with liquid. This cycle is a simple explanation for the dry surface that the *Huygens* probe encountered: The methane was taken up in clouds and at a later season will refill the basins.

Liquid ethane was also theorized to flow across Titan's surface, creating channels like rivers on Earth. The channels seen by the *Huygens* probe may also have been caused by cryovolcanism on Titan. Rather than erupting hot silicate or sulfur magmas, as volcanoes do on Earth and on Jupiter's moon Io, hypothetical volcanoes on Titan may erupt ammonia and water. Images taken by *Cassini* and *Huygens* indicate flow-shaped features that are consistent with ammonia geysers, and modeling by scientists shows that Titan has the correct compositions and temperatures to produce this kind of cryovolcanism.

Finally in 2007, another Cassini mission flyby was able to confirm the existence of lakes on Titan. In a study led by Ellen Stofan from Proxemy Research, radar images from the *Cassini* radar flyby of July 22, 2006, showed more than 75 lakes in the far northern latitudes of the moon. Though the surface temperature of Titan allows methane to exist as a liquid anywhere, everywhere except near the poles there is so little methane in the nitrogen-based atmosphere that any liquid methane would quickly evaporate into the atmosphere. The lakes occur in places where the temperature and pressure would allow both methane and ethane to be stable at the surface. The lakes lie in low points of the topography as they would be expected to and have channels leading to and from them. The *Huygens* probe and *Cassini*

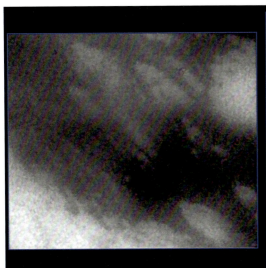 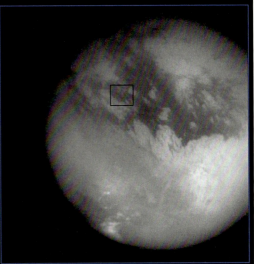

flybys have shown now that Titan does not have large oceans but instead a series of relatively small lakes. The lakes are estimated to cover about 1 to 4 percent of the planetary surface.

Shown above are two of the best images of Titan taken by the Cassini mission in 2004 previous to launching the *Huygens* probe. The wide-angle image of Titan has a scale of 6.2 miles (10 km) per pixel, while the narrow-angle image at left has a scale of just 0.5 miles (0.83 km) per pixel and covers about 249 miles (400 km). The surface has bright and dark markings with a streamlined pattern consistent with motion from a fluid, such as the atmosphere, moving from west to east (upper left to lower right).

A team of scientists from the University of Arizona, the U.S. Geological Survey, the German Center for Aerospace, NASA Ames Research Center, the California Institute of Technology, and Cornell University made a careful study of a particular lake, Ontario Lacus, in Titan's south polar region. Remote sensing from *Cassini* indicates that the lake contains methane, ethane, propane, and butane, an impressive soup of hydrocarbons, and an encouraging location in the search for life on other bodies. There will be much study of these fascinating lakes and streams and their links to rainfall and clouds on Titan. Titan is the first body outside the Earth known to have

Shown here are two images of the landing site of the Cassini-Huygens probe on Saturn's moon Titan. At right is an image showing most of Titan's disk with a scale of about six miles (10 km) per pixel; at left is an image of the landing site. (NASA/JPL/Space Science Institute)

this kind of cycle. On the Earth, water makes up the cycle of evaporation, condensation into clouds, precipitation, flow in rivers, and collection in lakes. On Titan, the methane, ethane, and other hydrocarbons not only make up the same cycle seen on Earth, they also form some of the bedrock and ground surface itself, as frozen ices.

Because of the liquid ethane and methane on Titan, the planet experiences rain from hydrocarbon clouds. Further study has revealed that the moon has an upper methane ice cloud layer and a lower nitrogen-methane cloud layer. These lower clouds are thought to produce a near-constant drizzle, wetting the whole surface of the moon. Rain that reaches the surface of a planet or moon is an unusual phenomenon in the solar system: there is none on Mercury, little or none on Venus, none on the Moon, and none on Mars. Titan's low gravity would have as much effect on the character of its possible rain as its strange compositions would. Titan's raindrops would be much larger than those on Earth, perhaps the size of grapes, and they would fall very slowly, more like large snowflakes on a calm day on Earth.

During Titan's year, Stofan and her colleagues suggest that lakes in the winter season will expand by the steady rain of methane. Lakes in the summer hemisphere will shrink through evaporation. Titan's midlatitudes also show erosional channels that were likely made by flowing water, so the research team further suggests that periodically higher methane humidity in the equatorial regions will lead to violent methane thunderstorms and flash floods made of methane that would carve erosional channels.

The different seasonality and process of rainfall in the different latitudes of Titan may explain the different patterns of erosional channels seen on the moon. Channels at low- and midlatitudes can be as long as several hundred kilometers and are frequently not single channels but braided channels, multiple channels that combine and separate in complex patterns. These channels resemble desert washes on Earth and thus are thought by some *Cassini* team members to form as a result of short-term, violent flooding from methane rainstorms. Channels at high latitudes, where lakes are most often found, tend to be single, meandering channels that drain into or connect lakes. A few channels are particularly wide (up to two miles [three km]) and deep (up to several hundred yards or meters) and seem to be cutting down deeply into Titan's crust, much as big canyons do on Earth. Though the shapes of channels are giving insight into surface processes on the moon, only about 1 percent of Titan's surface has channels. About 95 percent of the moon's surface has neither lakes nor channels.

Because winds are generated on planets by uneven heating from the Sun and the Sun is so far from Titan, winds should be mild. Some wind on Titan may also be created by Saturn's gravitational tides. Even mild winds, though, can create waves on a lake. Because the gravity of Titan is so low compared to the Earth and winds are low at high latitudes, waves on Titan are expected to be only about four inches (six cm) high with a wavelength of less than three feet (one m).

Waves in the lake basins can be expected to erode their shores, just as waves on Earth do. Titan's waves, though, may have a different and effective means of erosion, beyond simply the force and friction of the liquid striking rock and sand: bubble cavitation. When bubbles in a liquid pop when in close contact with a solid surface, they place a great deal of force on the solid surface, albeit over a very small area. The physics of this phenomenon, cavitation, are not well understood, but it has become very important to the military. Submarines can be detected underwater by the loud pops of bubbles cavitating next to their propellers, and this cavitation also damages propellers, prompting engineers to try to develop propellers that cause less cavitation. On Titan, cavitation may even be a problem in waves, because the ethane/methane liquid oceans are very volatile (liable to form a gas rather than a liquid) and so even a slight decrease in pressure, as could be found in parts of a breaking wave, could cause the liquid to form gas bubbles.

This radar image of the surface of Titan was acquired on October 26, 2004, when the *Cassini* spacecraft flew within about 745 miles (1,200 km) of the surface. A large dark circular feature is seen at the western (left) end of the image, but very few features resembling fresh impact craters are seen, suggesting that the surface is relatively young. The image is about 93 miles (150 km) wide and 155 miles (250 km) long, and is centered at 50 degrees north latitude and 82 degrees west longitude. The smallest details seen on the image are about 1,000 feet (300 m) across.

In 2005 and 2007, *Cassini* made flybys of Titan specifically to make radar maps of the moon's surface. The channels from the *Huygens* probe are a mild kind of topography, but the *Cassini* team led by Jani Radebaugh of Brigham Young University found high topography they are calling mountains. These mountains have elevations as high as 6,300 ft (1,930 m) and slopes as steep as 37 degrees. In relation to the two bodies' radii, these mountains on Titan would be the equivalent of peaks of 15,826 feet (4,825 m) on Earth. These are high mountains, though on Earth there are well over 100 mountains taller than 15,826 feet. (As a note, mountains can be measured by their height above sea level or by a measurement called prominence, which is the elevation of a summit relative to the highest point to which one must descend before reascending to a higher summit.)

Scientists do not yet know how these mountains were made. The processes that are possible on Titan are similar to many found on Earth: Compressive forces in the crust that

Titan's surface was imaged by Cassini prior to the landing of the Huygens probe. Huygens itself sent back images of a channeled landscape that had clearly supported flowing and pooling liquids. (NASA/JPL)

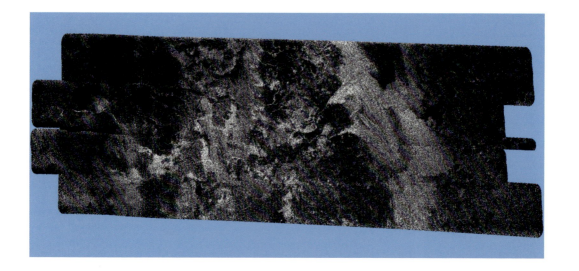

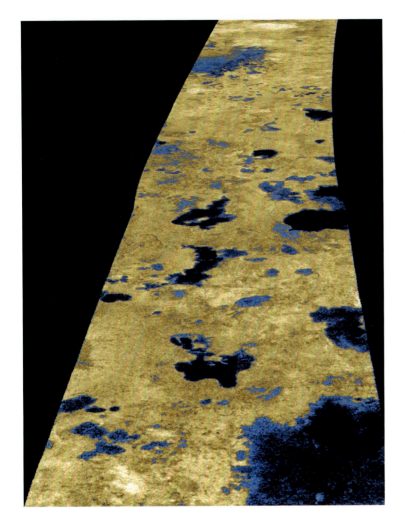

Low-backscatter radar regions in this image from Cassini *are colored blue, indicating liquid lakes on the surface of the moon Titan.* (NASA/ JPL/USGS)

push mountains up; extensional forces that allow blocks in the crust to drop down while neighboring blocks remain high; meteorite impacts that produce high crater rims; or erosion around a high-standing mesa.

There are many important similarities between Titan and the terrestrial planets: Titan's atmosphere is similar to the early Earth's; its surface pressure is higher than the Earth's by about half; it is larger than the Moon but smaller than Mars; and Titan appears to have had liquid oceans and may again in the future. Though scientists are hesitant to suggest it, it is possible that Titan could be hospitable for some form of life.

Like the moon Europa, there is some vanishingly thin hope that life might be found on this strange moon.

23. Hyperion

Hyperion is named for one of the Titans of Greek mythology. Hyperion and Theia, another Titan, produced Helios (the Sun), Selene (the Moon), and Eos (the dawn). Hyperion is an elongated satellite with a chaotic spin discovered separately in 1848 by both William Lassell, an amateur English astronomer and professional brewer, and William C. Bond, an American watchmaker and amateur astronomer, with the help of his son George P. Bond. Hyperion measures 230 × 174 × 140 miles (370 × 280 × 226 km), and orbits at average of 920,275 miles (1,481,000 km) from Saturn. Hyperion is the largest irregular moon in the solar system. Its surface is dark (0.3 albedo), similar to Iapetus's and Phoebe's surfaces but differing from all the moons interior to itself.

Hyperion is thought to have once been a larger object, almost destroyed by impacts. Its orbit is such that the detritus from the impacts would likely have hit Rhea, perhaps explaining Rhea's heavily cratered surface. Hyperion does not revolve neatly on its axis, as other moons do, but tumbles along its orbit. *Voyager* data indicated that even its tumble is not regular, that Hyperion's period of rotation varied from one orbit to the next. Newer data shows that its rotation period is more regular, close to 13 days. Tumbling would be a natural outcome of catastrophic impacts, lending support to that hypothesis, though other astronomers believe Hyperion is forced to tumble by gravitational interactions with massive Titan.

In addition to its many craters and pits, Hyperion's surface has a scarp named Bond-Lassell Dorsum that is 120 miles (200 km) long. This scarp may have been caused by movement along faults initiated by shrinkage of the moon during cooling, or it may be a suture zone from the impacts that nearly destroyed the moon.

24. Iapetus

In ancient Greek mythology, Iapetus was one of the Titans, a child of Gaia. He was himself the father of Atlas, among others. Iapetus the moon was discovered by Giovanni Cassini in 1671. It is 900 miles (1,440 km) in diameter, and orbits at a distance of 2,226,000 miles (3,561,000 km) from Saturn.

Unlike any other planetary body in the solar system, Iapetus's leading hemisphere is dark, and the other light. The dark leading hemisphere has an albedo of about 0.05, while the bright trailing hemisphere is 10 times brighter. Cassini himself noted this and wrote that he could see Iapetus on one side of Saturn but not on the other. His hypothesis is still the leading theory: Dust from the moon Phoebe, thrown off by impacts, is swept up by Iapetus as it revolves around Saturn. The light hemisphere is covered with ice relatively untouched by dust, and on the frontal hemisphere, methane ices may react with dust to create even darker substances.

All the moons exterior to Iapetus, excluding Phoebe, have been discovered since the year 2000. Their naming schemes changed from the traditional Roman or Greek. All retrograde moons have been named after figures from ancient Norse mythology, including Ymir, Suttung, Mundilfari, Skadi, and Thrym. A group of moons with orbits at about 34 degrees inclination, including Tarvos, Albiorix, and Erriapo, have been named after figures from ancient Gaulish religion. A third group of moons with orbits at about 46 degrees inclination, including Ijiraq, Kiviuq, and Paaliaq, were named after figures from traditional Inuit religion.

25, 26. Kiviuq and Ijiraq

Kiviuq is named for a hero in a traditional Inuit religion. Ijiraq was a spirit specializing in hide-and-seek, hiding children so well they were never found. In August 2000, Brett Gladman, an astronomer from the University of British Columbia, and a team of collaborators including J. J. Kavelaars, Jean-Marie Petit, Hans Scholl, Matthew Holman, Brian G. Marsden, Phil Nicholson, and J. A. Burns, discovered Kiviuq, Ijiraq, and Paaliaq with the 3.6-m Canada-France-Hawaii telescope on Mauna Kea, Hawaii. They were the first satellites of Saturn discovered since the Voyager 2 mission in 1981. These moons, along with Paaliaq and Siarnaq, all have orbits with about 46 degrees inclination.

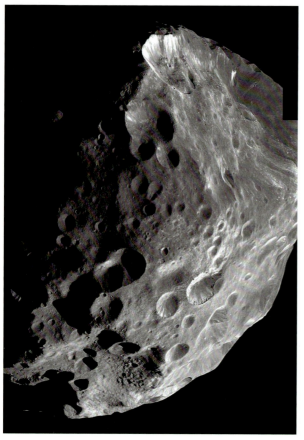

Saturn's moon Phoebe shows bright streaks on crater slopes that might be revealed by the collapse of overlying darker material from the crater wall. Aside from its bright streaks, Phoebe is one of the darkest bodies in the solar system and may be an icy Kuiper belt object captured by Saturn. (NASA/JPL/Space Science Institute)

27. Phoebe

Phoebe is named for one of the Titans in ancient Greek mythology. She was a goddess of wise counsel and mother of Apollo. Photos from the Cassini-Huygens mission show Phoebe's highly cratered and irregular surface. There are icy streaks within the crater walls, indicating that Phoebe may be an icy object with a dark mantle of dust and other material. As the moon has a retrograde orbit, it is almost certainly a captured asteroid. Its high ice content may indicate that it originated far out in the solar system, and thus may be the first object from the Kuiper belt to be observed closely by a space mission (not even Pluto has ever been visited).

Bright streaks on the rim of the large crater in the north of the image of Phoebe shown above may have been revealed by the collapse of overlying darker material from the crater wall. The large crater below and right of center shows evidence of layered deposits of alternating bright and dark material. Parts of Phoebe's irregular topography can be seen in and near the shadows at the lower left and upper left parts of the image, possibly crater rims or mountain peaks that are just struck by the first light of sunrise.

28. Paaliaq

Like Kiviuq and Ijiraq, Paaliaq is named for a giant in a traditional Inuit religion. In August 2000 Brett Gladman and a team of collaborators discovered Paaliaq along with Kiviuq and Ijuraq with the 3.6-m Canada-France-Hawaii telescope on Mauna Kea, Hawaii. Paaliaq is a member of the family of moons with orbits at about 46 degrees inclination.

29. Skadi (sometimes Skathi)

Skadi is named for a giantess in ancient Norse mythology who was the goddess of winter and the hunt. As they did with Paaliaq, Ijiraq, and Kiviuq, Brett Gladman and his collaborators discovered Skadi using the Canada-France-Hawaii telescope in Hawaii. The team discovered Skadi in September 2000, a month after discovering the three moons named for giants in Eskimo religion. Thrym and Mundilfari were discovered at the same time. Skadi is a member of the group of moons with retrograde orbits, all named for figures in Norse mythology.

Because many of Saturn's moons are very small bodies with almost nothing known about them, from Skadi outward in their orbits only a few selected moons will be discussed here. The others, as shown in the chart of Saturn's moons, are all small irregular satellites and appear even in good images simply as points of light.

30, 33, 34, 36. Albiorix, Erriapo, Siarnaq, and Tarvos

The Brett Gladman team discovered these four moons in late September 2000 using the Canada-France-Hawaii telescope. Albiorix, Erriapo, and Tarvos are named for figures from ancient Gaulish religion and Siarnaq for a figure from a traditional Inuit religion. Tarvos was a giant bull that often accompanied the master god Teutates. Albiorix was a high king. Erriapo was a mountain spirit. Siarnaq was the goddess of the sea and underworld. Siarnaq is a member of the Inuit-named moon family with orbital inclinations of about 46 degrees, while the other three are members of the family of moons with about 34 degrees inclination named for figures from Gaulish religion.

42. Mundilfari

Mundilfari is named for a giant in ancient Gaulish religion, the creator of the Moon and Sun. As they did with Paaliaq, Ijiraq, and Kiviuq, Brett Gladman and his collaborators discovered Mundilfari using the Canada-France-Hawaii telescope in Hawaii. The team discovered Mundilfari in September 2000, a month after the three moons named for Eskimo religion giants. Thrym and Skadi were discovered at the same time. Mundilfari is a member of the family of moons with orbits at about 46 degrees inclination.

47. Suttung (sometimes Suttungr)

Suttung is named for a fire giant in the Norse tradition. The moon was discovered by the Brett Gladman team. Suttung is a member of the group of moons with retrograde orbits.

53. Thrym (sometimes Thrymr)

Thrym is named for a frost giant in ancient Norse religion. Like Paaliaq, Ijiraq, and Kiviuq, Brett Gladman and his collaborators discovered Thrym using the Canada-France-Hawaii telescope in Hawaii. The team discovered Thrym in September 2000, a month after the three moons named for Eskimo religion giants. Mundilfari and Skadi were discovered at the same time. Thrym is a member of the group of moons with retrograde orbits, all named for figures in Norse mythology.

60. Ymir

Ymir is named for the first living creature and master giant in the ancient Norse religion. Ymir was an androgynous being with six heads who was formed from drops of icy chaos in Niflheim, near the beginning of time. Ymir was the grandfather of Odin and progenitor of mankind. The moon Ymir is a member of the group of moons with retrograde orbits, all named for figures in Norse mythology. The Brett Gladman team discovered Ymir, a tiny moon with diameter about 10 miles (16 km), in September 2000.

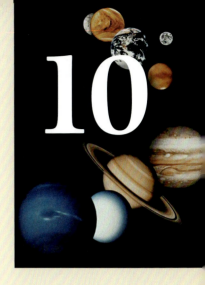

Missions to Saturn

Saturn was visited by three space missions before 2004: *Pioneer 11* and *Voyager 1* and *2*. These missions sent back images and data that greatly increased knowledge of Saturn and its processes and also showed new moons and rings. The new information from *Cassini-Huygens* allows scientists to check the hypotheses formed on the basis of the earlier missions' data, and its improved instrumentation and close approaches to moons generates new data and surprising images.

Pioneer 11 1973
American

Pioneer 11 was launched on April 5, 1973, and made its closest approach to Saturn on September 1, 1979. This was the first ever to approach Saturn. *Pioneer 11* was a flyby, with its closest approach at 12,990 miles (20,900 km) above Saturn's cloud tops, slightly closer than the later *Voyager 1* and *2* missions came to the planet. The mission passed through the ring plane twice and sent back 440 images. Two new moons and a new ring were discovered in these images. From the spacecraft's vantage point close to the planet, Saturn's rings appeared dark and the gaps between them bright, opposite to their appearance on Earth.

Voyager 1 1977
American

Voyager 1 made its closest approach to Saturn on November 12, 1980. Its visit was a flyby. The craft approached within 62,950 miles (101,300 km) of Saturn, passing between the planet's A and E rings. This course brought the craft within 45,000 miles (73,000 km) of Saturn's cloud tops. *Voyager 1* also passed near Titan, though it was unable to see through the moon's dense atmosphere, and then it proceeded on its way out of the solar system. *Voyager 1* measured the helium content of Saturn and found it to be considerably lower than Jupiter's (7 percent compared to 11 percent). The planets had been expected to have the same composition, so this finding was significant.

Voyager 2 1977
American

Voyager 2 made its closest approach to Saturn in August 1981, also flying within 62,950 miles (101,300 km) of the planet and passing between its rings. *Voyager 2* passed close to several satellites, then went on to Uranus. *Voyager 2*'s more sensitive cameras revealed many patterns in the planet's atmosphere that had been invisible to *Voyager 1,* and were, of course, still invisible from Earth. Both missions measured surprisingly fast winds, faster than those on Jupiter, and unexpected wind patterns. In addition to the east- and west-blowing winds, certain measurements indicate that Saturn has strong north-south winds at depth. *Voyager 2* measured temperatures and densities through the Saturnian atmosphere, and both missions measured auroras and rotation rates.

Cassini-Huygens 1997
American and European

Cassini-Huygens was launched in October 1997 as a joint effort of NASA, the Jet Propulsion Laboratory, the European Space Agency, and the Italian Space Agency. It arrived at Saturn on July 1, 2004. *Cassini-Huygens* carries an optical camera, a spectrometer, and radar, all for images and mapping.

Power is a major problem for a craft with such significant instrumentation far from the Sun where solar power cannot be used. *Cassini* is powered by three radioisotope thermoelectric generators that use heat from the decay of plutonium for power. The launch of the craft with 72 lbs (32.8 kg) of plutonium caused a lot of controversy; if the craft had exploded on takeoff a significant radiation pollution problem would have been created. NASA calculated that the chance of explosion in the first 3.5 minutes of flight was 1:1,400 and that health problems would be relatively minimal. Other scientists disputed these estimates. *Cassini* launched without incident, however, and its power source is reliable and will still produce over 600 watts of power even at the end of its current mission. However, a current plutonium shortage requires that future missions must find an alternative power source.

Cassini's principal objectives are to do the following:

* determine the three-dimensional structure and dynamical behavior of the rings
* determine the composition of the satellite surfaces and the geological history of each object
* determine the nature and origin of the dark material on Iapetus's leading hemisphere
* measure the three-dimensional structure and dynamical behavior of the magnetosphere
* study the dynamical behavior of Saturn's atmosphere at cloud level
* study the time variability of Titan's clouds and hazes
* characterize Titan's surface on a regional scale

On its way to Saturn from Earth, *Cassini* executed two gravity-assist flybys of Venus, then one each of the Earth and Jupiter to propel it to Saturn. Upon reaching Saturn, *Cassini* swung close to the planet, passing upward through a gap in the rings, to begin the first of ~70 orbits during the rest of its four-year mission.

Cassini made 44 close flybys of Titan, the first of which occurred in October 2004. On the third flyby, in January

2005, the 133-pound (300-kg) *Huygens* space probe parachuted onto Titan. The probe sampled the atmosphere and sent back photos of the surface. *Huygens* relayed data from near Titan's equator for several more minutes and became the most distant man-made lander in history. The probe beamed its data to the *Cassini* orbiter, which relayed the data back to Earth, where it was received by the Deep Space Network dishes at a rate of several gigabytes of data per day.

During the course of the *Cassini* orbiter's mission, it performed close flybys of bodies of interest, including more than 30 encounters of Titan and at least four of other icy satellites. *Cassini*'s penetrating radar allowed it to map Titan's surface in detail. The orbiter also made at least two dozen more distant flybys of the Saturnian moons. *Cassini*'s orbits will also allow it to study Saturn's polar regions in addition to the planet's equatorial zone.

In April 2008, NASA announced a two-year extension on the mission. In the extended mission, *Cassini* will focus on Enceladus and Titan.

Among *Cassini*'s accomplishments are the following:

* the discovery of four new moons of Saturn (Methone, Pallene, Polydeuces, and Daphnis)
* the discovery of a tiny new ring associated with the moon Atlas
* the first measurements of the temperatures of the rings
* a large storm at Jupiter's north pole
* a test that confirmed Einstein's theory of relativity to 20 parts in a million
* the first direct information about Titan's atmosphere and surface
* photographs of Enceladus's water geysers and astonishingly hot surface and interior
* discovery of lakes of methane and ethane on Titan

Cassini's extended mission will include 60 additional orbits of Saturn, including 26 more Titan flybys, seven Enceladus flybys, and one each of Dione, Rhea, and Helene.

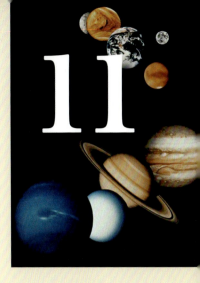

An Environment for Life? Titan, Enceladus, and Europa

Temperatures on the surfaces of planets and moons in Earth's solar system vary widely, from Neptune's moon Triton at -391°F (-235°C), to Venus under its insulating clouds at 860°F (460°C). Until recently life on Earth was thought to exist only within a relatively narrow range of conditions, determined by the behavior of water. Because every living thing on Earth has water in its cells, life was thought to exist only between the freezing point of water 32°F (0°C) and its boiling point 212°F (100°C).

REQUIREMENTS FOR LIFE

As scientists have worked to understand how life withstands extreme conditions, a special research topic has emerged: The study of forms of life that survive outside normal conditions, life-forms known collectively as extremophiles. Extremophile microorganisms have now been found that can live at temperatures as high as 266°F (130°C) and as low as 1.4°F (-17°C). Other extremophile microorganisms can survive extreme acidity or its opposite, basic fluids, from a pH of less than 0 (an acidity equivalent to battery acid) to a pH of 12 (a base slightly stronger than pure household cleaning ammonia). Still others

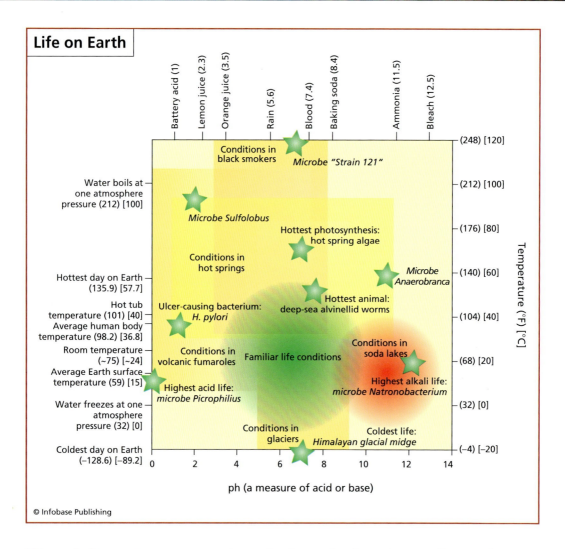

Life on Earth

This graph of temperature v. acidity (pH) shows the wide range of conditions under which life has been found on Earth and shows scientists how widely to search on other planets. (Modeled after Shock and Holland, Astrobiology 7, 2007)

can survive in fluids so salty they are crystallizing salt grains among the microorganisms, and others at pressures from the vacuum of space up to 2,000 atmospheres, that is, 2,000 times the surface pressure of Earth and 20 times the surface pressure of Venus. Some of the most extreme extremophiles are shown in the figure above. Discovery of these extremophiles has broadened the range of places scientists now search for life elsewhere in the solar system and the universe.

The Earth has had only primitive, single-celled life for almost 90 percent of the time there has been life on Earth. Even now, single-celled life-forms outnumber complex life-forms (including all the animals and plants we can see with our naked eyes) by many orders of magnitude. When envisioning life on another planet or moon, then scientists think it most likely that life there will be single-celled. When thinking about what kinds of planets and moons they should look at, they think about not only the conditions extremophiles have adapted to but also what the early Earth might have been like when life first evolved.

Atmospheric Composition of the Early Earth

The early Earth, by contrast to its current oxygen-rich atmosphere, probably had a nitrogen-rich atmosphere with additional carbon dioxide and methane. The best current hypothesis for the early Earth's atmosphere is that it had very little oxygen and that along with nitrogen, methane, and carbon dioxide, it contained hydrogen cyanide (HCN), phosphine gas (PH_3), ammonia (NH_3), and hydrogen sulfide (H_2S), all of which would combine to make a soup of organic chemicals that were ideal for the formation of life, though still entirely without oxygen.

In the young solar system, when these conditions were thought to exist on Earth, the Sun produced far less energy than it does now. Young stars require some time to get their internal reactors up to full production, and at the time life was emerging on Earth the Sun might only have produced 65 percent of the energy it does now.

The peculiar atmosphere that existed on Earth at that time helped to solve the problem of a cold Sun. Many of the atmospheric gases on the early Earth are so-called greenhouse gases that hold solar heat at the Earth's surface rather than letting it escape to space. Thus, the strange early atmosphere helped keep the early Earth hotter than it otherwise would have been. The higher temperatures speed chemical reactions and would have encouraged the development of life.

Planetary atmospheres are constantly bombarded with energy from the Sun, and when an energetic photon strikes a particle of the atmosphere that particle can be broken into smaller particles. This process is known as photodissociation or photolysis. In an atmosphere consisting of methane (CH_4), photodissociation produces acetylene (C_2H_2). Until recently acetylene was not thought to be useful for any microbial life. Acetylene was further thought to suppress life because it prevents a variety of chemical processes that oxygen-loving bacteria require to stay alive.

About 25 years ago, however, some species of bacteria were found that use acetylene for energy, without using any oxygen. Now a methane atmosphere with acetylene is considered a possibly hospitable environment for life that does not require oxygen, broadening the possibilities of life on early Earth and on other bodies with similar chemistries.

Water in Liquid Form

Life on Earth is impossible without liquid water. Water is used in the chemistry of cells, in the production of energy and waste removal, and also as an environment for life. Water is an excellent solvent, that is, most materials can dissolve into water and therefore be transported by water either inside an organism (as in blood or other fluids) or in an environment that the organism lives within.

At the limit of dryness on Earth, some single-celled organisms have been found that can go into a kind of hibernation called stasis for extremely long times, perhaps hundreds of years or even longer, keeping their processes going at very low levels. When these microorganisms in stasis again have liquid water, many revive quickly and begin again their chemical processes.

No life has been found that can exist entirely without water. Scientists think that other liquids might be used in the same way for alien kinds of life, but none has yet been found. Still, some fluid is thought to be needed for life to exist.

Organic Matter and Energy

Organic matter consists of molecules that contain carbon, often (but certainly not always) in combination with oxygen and hydrogen atoms. Life on Earth requires both organic matter and inorganic matter (molecules without carbon) to live. The organic matter is used to build the organism itself and sometimes for food. The inorganic matter can be used as a source of energy for primitive life such as some bacteria, as described in the next section. Organic matter is therefore a strong indicator of either life or the possibility of life on another body.

Plants and some other simpler life-forms on Earth now obtain the energy for growth and life through the process of photosynthesis, obtaining energy from sunlight. Photosynthesis and indeed all the chemistry of simple bacterial life involve taking an electron off one atom and using it to create a source of energy for the organism. Photosynthesis requires oxygen, and the early Earth appears to have had almost no free oxygen in its atmosphere. Therefore early life had to find other chemical processes to create energy for life.

Early life on Earth was likely of a kind called either chemoautotrophy or chemoorganitrophy. Chemoautotrophs obtain their energy by oxidizing inorganic molecules. "Oxidation" refers to the loss of an electron by an atom or a molecule. If an atom or a molecule has an extra electron, it is referred to as reduced. Chemoautotrophy requires two kinds of molecules to be available. First, reduced compounds such as HS^-, H_2, CH_4, or NH_3 that can give up an electron; and second, it requires ions or molecules such as Fe^{3+}, Mn^{4+}, SO_4^{2-}, or NO^{3-} that can oxidize (or take an electron from) the reduced compounds listed above.

The chemicals these early bacterial life-forms needed were probably scarce in their environment, and so scientists think these bacteria probably had a broad and complex ability to use many chemicals to survive in a difficult environment. These bacteria could not afford to specialize in just a few sources of energy.

Chemoautotrophs produce oxygen as a by-product of their other chemical reactions and thus may have helped raise oxygen in the atmosphere. They do not require oxygen to live, like the plants and animals we are most familiar with ("aerobic" life), and so are called "anaerobic" organisms. On Earth today chemoautotrophs make up the majority of organisms below 1,000 m in the oceans and beneath 500 m depth on land, where there is no sunlight available.

Heat in the Environment

Chemical reactions proceed more quickly at warm temperatures than at cold ones. From this simple fact comes the idea that some heat source is a great help for the development of life, in that organic molecules will make more reactions together and perhaps produce useful chemicals more quickly, and for the maintenance of life, as all the processes that maintain life happen faster and more easily at higher temperatures. At low temperatures chemical reactions proceed so slowly that life is inhibited; in laboratory experiments microorganisms will continue to grow and reproduce at lower temperatures but far more slowly. Below some temperatures, growth and reproduction stops and the organisms die, but this temperature is different for different organisms, and scientists do not know if there is an absolute lowest temperature beneath which life cannot exist.

Though heat helps chemical reactions proceed and therefore helps cells produce energy and grow and carry on the activities of life, at high temperatures chemical reactions can proceed so fast that cells cannot extract energy before the reaction has moved on. Scientists suggest, from this idea, that there is some temperature above which life could not exist, even if the cells had developed a method to survive with water above its boiling point. However, not even a theory exists for what this highest possible temperature might be.

Planets and moons have a variety of possible heat sources. The first for the Earth and Mars is the Sun, pouring its energy into the atmosphere. The second is heat being released from the planet's interior from both the original heat of formation and from decay of radioactive elements. This second heat source is not constant over the age of the planet but declines with time. A third possibility is heat produced by tides in the planet or moon from a larger body. The Galilean satellites of Jupiter (Io, Ganymede, Europa, and Callisto), for example, are heated significantly by internal friction produced by large gravitational and magnetic tides from Jupiter. Io, for example, is heated so greatly by tides from Jupiter that it experiences near-continuous volcanic eruptions from its tidally melted interior.

How Common Are These Conditions in the Formation of Planets?

First, scientists look for Earthlike planets in the hopes that familiar kinds of life may have developed on them. Earthlike planets have metallic cores that consist mainly of iron, surrounded by a silica-based rock mantle and surfaced with a silica-based rock crust. Even more important, the Earth has an insulating atmosphere and is at the right distance from the Sun to have surface conditions that allow liquid water to exist. If the Earth were closer to the Sun, water would be vaporized into steam, and if it were farther from the Sun or had a thinner atmosphere (like Mars), water would be frozen.

Planets that have water that is mainly frozen but occasionally thawed to liquid, for example during the planet's summertime, may also be reasonable hosts for life. As on Earth where bacteria have been found to go into stasis when water

Liquid water is necessary for all forms of life yet found. The temperatures necessary for liquid water are present or nearly present on many solar system bodies.

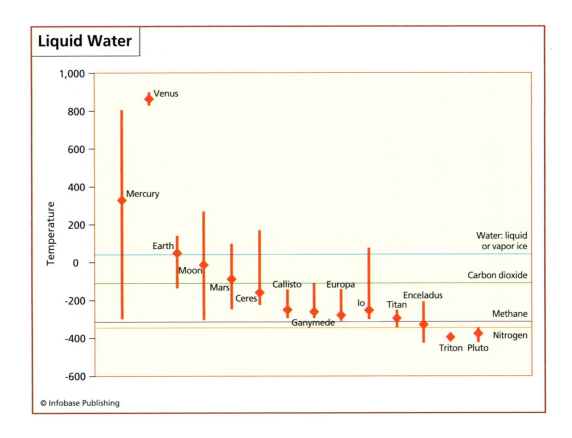

was unavailable and then revive when water returned, so might life on these planets thrive in the summer and go into stasis in the winter.

Mars as the Early Earth Analog

Mars, one of the Earth's closest neighbors, is an Earthlike planet and a natural target in the search for life. Observers have thought they could see signs of life on Mars for centuries and were particularly outspoken about this possibility in the 1800s, when canals, buildings, and forests were thought to be seen. Though all of these have since been proven to be optical illusions, the search for life on Mars continues to the present moment with the *Mars Phoenix Lander* and its measurements of the composition of the Martian soil.

Mars has frozen water in its soil, carbon dioxide ices, and apparent liquid seepage between rock and soil layers that leaks out of cliffs and within impact craters. Though the average surface temperature is -85°F (-65°C), the planet's interior heat means that liquid water or water mixtures may be stable at depth in the soil and rock. Mars is therefore a colder and drier version of the Earth, but similar in many other ways, and so the hope that microbial life, either alive today or as fossil remains, may still be found on that planet.

Super Earths in Other Solar Systems

A great topic of interest is the possibility of finding life in planetary systems around other stars. Techniques for detecting planets around other stars (exoplanets) are improving continuously, and instruments are being launched on satellites to take advantage of improved observing conditions outside the Earth's atmosphere.

Most of the more than 200 exoplanets that have been discovered are giant gas planets similar to Jupiter. These are not great candidates for life, or at least, because of their very dense and thick gas, any life that had adapted to those conditions would be very difficult to detect.

Scientists are working hard to detect planets more like the Earth, planets with rocky mantles and the relatively thin atmospheres of Earth and Venus (thin in comparison to the gas envelopes of Jupiter or Neptune). Planets the size of Earth are still too small to be seen using even the best current observations, but scientists are finding more and more of the planets described as super Earths. These are Earthlike (rocky) planets of up to 10 times the mass of Earth, and they are candidates to contain life.

Several super Earths have been found since 2007 that may exist at the right distance from their stars to have liquid water, one of the prime targets for life. Perhaps the most famous of these planets is Gliese 581c, at about five Earth masses. The next challenge is to find a way of detecting life at such a great distance from the Earth.

Icy Moons in Our Solar System

Because the Sun contains about 99.8 percent of the mass of the total solar system, scientists commonly assume that the bulk composition of the Sun is a good approximation for the bulk composition of the whole solar system. The composition of the Sun indicates that in the cold outer solar system planets and their moons will form largely from solid ices, and that these ices will be mostly water (H_2O), with smaller amounts of ammonia (NH_3), methane (CH_4), carbon dioxide (CO_2), and nitrogen (N_2).

The solid parts of the outer planets and moons are expected to consist of these ices with a similar or lesser amount of silicate rocks, like the mantle of the Earth. The moons appear indeed to be made of ices with rocky cores, with the exception of Jupiter's moon Io, which is a rocky, volcanic body with no ice. These bulk compositions are encouraging as a starting point for life, since they include water and the important combination of inorganic and organic chemicals. All that appears to be missing is heat.

The icy moons have surface temperatures from -350°F (-210°C) to -210°F (-130°C), temperatures low enough to slow or stop the chemical reactions needed for life. What energy the Sun fails to deliver at those great distances may be replaced, at least for the interiods of the moons of the large planets, with frictional heat from gravitational and magnetic tides.

The moons Titan, Enceladus, and Europa, discussed here, may be the most likely places for life to exist in our solar system away from Earth. Mars and the moons Ganymede, Callisto, and Triton are also possible locations.

Each of these bodies either has an atmosphere sufficient to prevent all gases or liquids from evaporating into the vacuum of space or an internal ocean. They have liquid water or another liquid that might support a form of life that might be recognizable.

Detecting Life on Other Planets and Moons

Many scientists are attempting to predict what can be measured to show the existence of life on another planet. Missions like the Mars Phoenix Lander, which is currently looking for signs of life on Mars, are limited in where on the planet they can look (near their landing place and only with the near-surface atmosphere and soil). Scientists are trying to predict what atmospheric compositions indicate life, since these can be measured from a distance and do not require landers. The atmospheric compositions of planets in other solar systems may soon also be measured.

The atmospheric compositions that indicate life are called biosignatures. Right now, since the Earth is the only body known to have life, the only biosignatures known to scientists are those from Earth. The challenge to researchers is to imagine kinds of life that do not exist on Earth and then what kinds of trace gases in the atmosphere they might produce.

Oxygen is a highly reactive element. The persistence of oxygen in Earth's atmosphere is due to the continued work of organisms that produce oxygen; otherwise all or almost all oxygen would react with Earth's surface materials and other atmospheric constituents and shortly the atmosphere would be devoid of oxygen (for example, oxygen reacts with iron to form rust; in this reaction the oxygen is bonded to an iron atom and removed from the atmosphere). Oxygen in an atmosphere is therefore scientists' primary indicator of life on a planet.

The instability of oxygen in the atmosphere and its requirement of constant replacement by life is an example of disequilibrium thermodynamics, meaning processes that have not completed all reaction to a final resting state with no further reaction. On the Earth, life is constantly creating chemical reactions that give off oxygen, never allowing all the oxygen to leave the atmosphere through reaction with surface materials.

By analogy to the necessity of life to maintain oxygen, when looking for life on other planets scientists seek other molecules in atmospheres that would not remain without a constant source. These researchers are looking for other kinds of disequilibrium thermodynamics in which atmospheric gases are produced continuously and therefore exist continually in the atmosphere, when otherwise they would react with other chemicals and disappear.

The surface geology of a planet may also provide clues about life. On Earth limestone (calcium carbonate, or $CaCO_3$) can be precipitated, or crystallized, directly from seawater. The surface rocks of Earth, however, contain vast amounts of limestone, far more than can be explained by production directly from the oceans. Life-forms also produce limestone in the form of shells, such as seashells and shells from plankton, microscopic organisms that exist in the oceans in huge numbers. Most of the limestone on Earth has been produced by life. Therefore the amount of limestone on Earth is a biosignature: It indicates the existence of life.

The rate of erosion on Earth is also a biosignature. Much of the weathering and breakdown of rock into clay minerals and sand is done by life and not simply by frost and rain. Lichens, for example, eat into and digest rocks. Though each lichen is a small and easily overlooked organism, together all the lichens on Earth are estimated to weigh 13×10^{13} tons, greater than all the biomass in all the oceans. Thus their action in weathering rocks is substantial, and the rocks on Earth's surface would be far less weathered without them.

Researchers face the challenge of trying to imagine what entirely alien life might be like. Complex intelligent life like man or like large mammals would be easy to detect, we hope, because they would sufficiently alter their environment with paths and roads and perhaps buildings or even technology emitting energy waves. Based on the kinds and numbers of life on Earth, though, it seems likely that life on another planet is most probably simple microscopic life, like bacteria on Earth. If we are fortunate, this life might use sources of energy similar to those of simple life on Earth, that is, chemical reactions that involve removing an electron from one molecule and adding it to another. What other means of survival exist in the universe and how would we detect it? Trying to answer this almost unanswerable question is the job of many researchers today.

TITAN

Titan probably has many similarities to the early Earth and is therefore an especially interesting target in the search for life. Titan's similarities to the Earth include atmospheric composition and the existence of a hydrologic cycle, that is, liquid that is collected in surface lakes, evaporated into clouds, and then rained back onto the surface. On Titan, however, the cycle is not made from water but from methane.

Titan also has an ocean, likely made of water, but its ocean exists inside the moon between layers of ices and not on the moon's surface.

Atmospheric Composition

Titan's atmosphere is made of primarily nitrogen (95 percent). Titan is one of the very few bodies in the solar system with a nitrogen-dominated atmosphere, and the Earth is the only planet with a nitrogen-dominated atmosphere.

Titan's atmosphere is thought to be very like the early Earth's, containing methane and nitrogen and a wide range of other organic compounds but not oxygen. While Earth's early atmosphere was likely warm (which speeds chemical reactions) and constantly added to by volcanic eruptions, Titan's atmosphere is cold and less changeable than the early Earth's is thought to have been.

Acetylene is known to exist in the atmosphere of Titan and is thought to be produced by photodissociation of methane, a major constituent of Titan's atmosphere. Early Earth is thought to have had a methane atmosphere, as Titan has today. The discovery of bacteria on Earth that use acetylene for energy and may have existed since the early Earth times during which methane and acetylene were more common in the atmosphere than was oxygen raise the possibility of life on methane-acetylene worlds. The presence of acetylene in the atmosphere of Titan therefore means that there is a possibility of anaerobic life on Titan, living at the surface and creating energy from the atmosphere.

Liquid Surface Lakes

Earth has a complete cycle based on water, in which liquid water collects in lakes and oceans on the surface and then is evaporated into clouds and later condenses into rain that falls back to the surface.

Titan has a similar cycle, but instead of water it is made from methane. Titan has methane clouds from which large droplets of liquid methane fall and collect into liquid methane lakes on its surface. Clouds appear in bursts at Titan's south pole and in streaks elsewhere in the southern atmosphere. The moon's surface has structures that are almost certainly river valleys and lakebeds, filled during a rainy season and otherwise dry.

The surface liquid on Titan for many years had been suggested to exist based on theory, but only the Cassini mission was able to confirm its existence, during flybys in 2007. In a study led by Ellen Stofan from Proxemy Research, radar images from the *Cassini* radar flyby of July 22, 2006, showed more than 75 lakes in the far northern latitudes of the moon. Though the surface temperature of Titan allows methane to exist as a liquid anywhere, everywhere except near the poles there is so little methane in the nitrogen-based atmosphere that any liquid methane would quickly evaporate into the atmosphere.

The lakes occur in places where the temperature and pressure would allow both methane and ethane to be stable at the surface. The lakes lie in low points of the topography, as they would be expected to, and have channels leading to and from them.

Some researchers have suggested that these cold lakes may hold bacteria that live on methane. These bacteria might also live on acetylene, like those hypothesized for the atmosphere, since acetylene can dissolve into liquid methane in the lakes.

All of the vast changes that have occurred on Earth since life first arose mean that observing and understanding the conditions of early life of Earth is very difficult. Titan, however, may show scientists what that early Earth was like in many ways. By studying Titan, scientists are looking at something like the prebiotic (before life) Earth and may learn more about how life arose on our planet.

Enceladus's icy surface may hide inner liquid water regions. (NASA/ JPL/Space Science Institute)

ENCELADUS

One of the Cassini mission's biggest discoveries was plumes of liquid water erupting through the icy crust of Enceladus and flying as high as 50 miles (80 km) above the surface. These

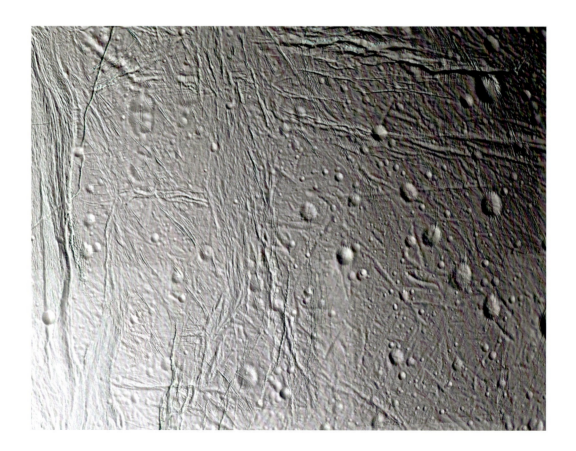

Enceladus's water jets become visible when it lies between Cassini and the Sun. (NASA/JPL/ Space Science Institute)

plumes may be caused by the gravitational field of Saturn, which slightly deforms Exceladus through tidal stresses. This slight deformation may squeeze internal reservoirs of water and cause some eruption onto the surface.

In 2008, *Cassini* flew just 30 miles (48 km) above Enceladus and took the highest-resolution photos yet obtained. These photos show that the plumes originate in cone-shaped pits up to 1,000 feet (300 m) deep. Mounded deposits near the cones are thought by mission scientists to be piles of snow left by the geysers as they freeze.

The plumes originate at warm plots near Enceladus's south pole. Enceladus's average surface temperature is about -324°F (-200°C), and parts of the planet can be as cold as -400°F (-240°C). In the warm spots the temperature rises to an average of -306°F (-188°C) and in places as high as -207°F (-133°C), more than 115°F (65°C) warmer than the Moon's average temperature.

But how does liquid water exist at these extremely low temperatures? Pure water would be frozen solid long before the temperatures reached -207°F (-133°C), but water mixed with ammonia can remain liquid at far lower temperatures, as low as -142°F (-97°C). This lowest temperature at which water and ammonia together can remain as a liquid is still warmer than any temperature measured on Enceladus, but these readings on Enceladus were measured at the surface—could the moon be warmer inside, where the liquid is originating? This is the scientists' hypothesis.

Though ammonia is the best candidate for lowering the melting point of water, only tiny traces of ammonia have been detected in the plumes. While the icy crust of Enceladus is almost pure water, the plumes appear to be made primarily of water and nitrogen but also have carbon dioxide and carbon monoxide and may have methane, acetylene, and propane. Thus Enceladus's internal water ocean may be filled with hydrocarbons, good sources of energy and growth for life.

Even this hypothesis cannot explain the fundamental question: What causes the heat in Enceladus in the first place? Theoretical calculations indicate that Enceladus does not receive enough tidal friction from Saturn to explain even a fraction of the heat seen in the moon. There may be an especially enriched region of radioactive elements in the south pole of Enceladus, but there is no simple explanation for why these elements would be there. On Earth, at least, truly unusual circumstances are required to make radiogenically enriched regions like what would be needed to heat Enceladus. So the heat on Enceladus that melts the liquid and forms the geysers remains a mystery.

These hydrocarbons may be from the breakdown of ices or rocks in the interior or may exist in liquid form. An internal water ocean in contact with rocks is hopeful for the existence of microbial life, since on Earth bacteria use nutrients both from rocks and liquids.

EUROPA

Ridges, troughs, bands, and other linear structures on the surface of Europa are interpreted by scientists as fractures in an ice shell that have filled again with liquids that froze. An

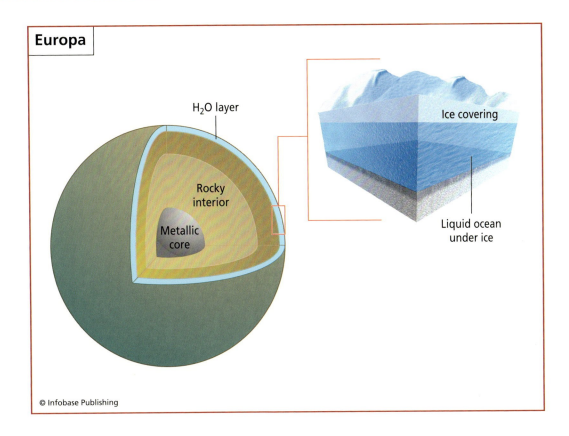

Europa

H₂O layer

Rocky interior

Metallic core

Ice covering

Liquid ocean under ice

© Infobase Publishing

interpretation of these structures is that motion inside Europa breaks the moon's crust and that upwelling liquid then fills and freezes in the fractures. Both the apparently frequent movements in the crust and the availability of new ice to fill them imply that Europa's interior is at least partially liquid. Though this hypothesis states that an interior liquid ocean wells up into cracks, there is no evidence of flooding onto the surface. Perhaps the exceptional cold on Europa's surface, averaging -274°F (-170°C), is sufficient to instantly freeze a crust on any upwelling liquid and prevent surface flows.

According to an alternative hypothesis, Europa does not have an internal reservoir of liquid but instead the fracturing process pulverizes the icy crust such that it looks like fresh ice but is not. Most scientists, however, are confident that Europa does indeed have an internal ocean with a complete frozen crustal layer. Data on Europa's spin indicate a liquid layer and

Europa is thought to have an interior water ocean, making it a special target in the search for life.

suggest that the icy shell capping the moon's ocean is between 45 miles (75 km) and 90 miles (150 km) thick.

By comparing the numbers of craters on Europa's surface with the expected cratering rate in that part of the solar system, an estimate for the age of Europa's surface can be made. Europa's surface appears to be 30 to 70 million years old, young even by Earth standards, and Earth has one of the most quickly renewed, freshest surfaces in the solar system. By comparison, much of Mars's surface is thought to be billions of years old. Thus Europa is known to be active currently, with a relatively thin moving ice crust being resurfaced.

Remote sensing of the moon shows that the moon's internal ocean consists mainly of frozen water, with a very small amount of carbon dioxide and some carbon-, hydrogen-, sulfur-, and oxygen-bearing molecules. Europa's internal ocean is likely to have a volume of water equivalent to many Earth oceans.

The limiting factor for life on Europa may be a lack of chemical energy. There may be no way for the internal ocean to remain out of equilibrium chemically, that is, to have reduced chemicals and oxidizing chemicals that chemoautotrophs can use to produce chemical energy. If some connection to the surface is maintained, either through geysers or occasional meteorite strikes or through some other process, then the possibility of chemical energy still exists. Without some process connecting the ocean to the surface, the chemicals in the isolated interior ocean will complete all possible reactions and come to equilibrium, leaving no energy for life.

QUESTIONS FOR FUTURE RESEARCH

Life is the central fact of our existence and the nature of our planet. It remains to be understood on the most fundamental bases. Some of the first-order questions under current research include:

* ✹ How frequently and for how long do habitable conditions appear on a planet?

* What conditions in terms of planetary mass, accretion dynamics, impact history, and interactions among the star and other planets are necessary to form habitable conditions?
* How persistent and extensive must habitable conditions be to sustain life over the longer term?
* Must habitable conditions exist at the surface, or are habitable conditions at depth sufficient?
* How common are such planets?

So much about life is not understood. The biggest question of all, what is life itself, remains unanswered. Of the questions that surround this particular one, others remain unanswered. These areas of research are fundamental to understanding life on this planet, the Earth itself, and its place in the universe.

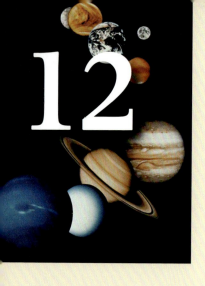

12

Conclusions: The Known and the Unknown

Though both Jupiter and Saturn have been the objects of human attention since prehistory, and of human study in a modern sense since at least the 16th century, more questions remain than answers. These planets are so unlike the Earth that in their study the theory of planets needed to be rewritten, starting from the accretion of the original planet and continuing through today's internal processes, weather, and the interactions of magnetic fields, moons, and rings. The sheer mass of the systems makes them unlike the Earth and the other terrestrial planets: Several of the moons of these planets are of a size similar to or larger than the small terrestrial planets, and the gas giants themselves reach internal pressures high enough to force hydrogen and helium into a metallic phase. The size and number of moons makes them particularly interesting, and relatively little is known about them.

1. **What are the surface features and compositions of the satellites?**
   A number of the larger moons of Saturn and Jupiter are thought to have formed along with their planets early in solar system history, from the evidence of their round

orbits that lie in the planets' equatorial planes, and in part from their compositions. Neighboring moons to a single planet, though, appear to differ completely in composition (for example, Io is a silicate moon and Europa is an icy moon) and in geologic activity. Because their surface features, geologic composition, and internal processes are not well understood, their formation and relationship with their planets are similarly hazy.

2. **How are moons and rings connected, in terms of development and maintenance, and also in terms of composition?**

   Jupiter's rings appear to consist almost entirely of dust, while Saturn's consist primarily of ices. Though their formation is not well understood, rings are thought to be transient phenomena in terms of solar system time scales. Why, then, do all the gas giant planets now have rings, and why are their rings different in shape and composition? The complexities of Saturn's rings make clear that some moons maintain clear spaces between rings, others form sharp edges along rings, others perturb rings into knots and braids, and still others create rings by shedding material into their orbits. Rings may therefore be formed by moons already orbiting the planet in question, and theory holds that they may also be formed by captured and destroyed comets or other bodies. The dynamics of ring formation and destruction may also shed light on processes in the early solar nebula as the planet themselves were forming.

3. **What are the structure and composition of Jupiter's and Saturn's interiors?**

   Though there are various competing theories for the formation of giant planets, none fully explains the size, exterior composition, and position in the solar system of these icy giants. Jupiter and Saturn may have rocky cores, or they may not. Their cores are only constrained to be between one and about 10 Earth masses, not a narrow margin. The two planets must consist mainly of hydrogen and helium, based on both solar abundances and the size and density of the planets, but their internal structures and dynamics remain a realm of theory and not of measurement.

4. **Why is Io so volcanically active?**

Though Io's volcanism is usually attributed to tidal heating from the intense friction and flexure caused by Jupiter's gravity field, calculations of the heat that could be caused by tidal frication may not equal what Io is seen to release through volcanic activity. None of the hypotheses for heating Io, including electric heating from the dynamic effect of Jupiter's magnetic field, and radiogenic heating from unstable isotopes in Io, may be sufficient to explain the very high heat flow of the planet. Beyond the heat problem, Io has been volcanically active at high volumes for millennia. This constant volcanic activity has the inevitable result of reducing the moon's budget of elements that melt at low temperatures through loss to Jupiter's magnetic field from volcanic plumes. If the moon has lost its low-temperature component, then it would require higher and high temperatures to continue melting and producing volcanic activity. Its energy requirements go up, therefore, with time.

5. **Why does Ganymede, and some of the other moons, have their own magnetic fields?**

Planets are thought to produce magnetic fields through liquid convection around a solid center, such as the speedy convection in the Earth's outer molten iron core that surrounds its inner solid core. Ganymede should have long since cooled in its interior past the point that it could create a convective dynamo. The friction of tidal forces with Io, Europa, and Jupiter itself may contribute to a magnetic field, but its source is currently unknown. Titan and Io, silicate moons, and Europa and Callisto, ice moons, also produce their own magnetic fields.

The excitement of continuous discovery in these planetary systems continues. Both Jupiter and Saturn are immensely complicated systems not unlike miniature solar systems on their own: Jupiter has 63 known moons and its own ring system. Saturn is famous for its rings, and it also has 62 known moons. Jupiter is the largest planet in the solar system, but it contains only one-tenth of 1 percent of the mass of the Sun. If it had been only 80 times larger, it would have had the mass required to begin nuclear fusion in its interior and the solar system would have had two stars. Saturn has the youngest and freshest ring system in the solar system, presenting a beautiful laboratory for the study and understanding of rings. Each of these gas giants has a moon that has almost come to eclipse the scientific interest of the body itself: Saturn's largest moon, Titan, has an atmosphere like that of early Earth, and Jupiter's moon Europa has a water ocean and thus is the most likely place in the solar system to find life now (Mars may have had life in the past, but is less likely to have life now). Io, one of Jupiter's large moons, is the most volcanically active body in the solar system, far more so than Earth itself (and the only body aside from Earth on which man has witnessed volcanic activity) and provides a model for the massive flood basalts that have disrupted Earth's climate in the past. Through these moons, and many more unusual aspects of these giant planetary systems, Earth's past, its present, and possibly its future can be studied.

APPENDIX 1:

Units and Measurements

FUNDAMENTAL UNITS

The system of measurements most commonly used in science is called both the SI (for Système International d'Unités) and the International System of Units (it is also sometimes called the MKS system). The SI system is based upon the metric units meter (abbreviated m), kilogram (kg), second (sec), kelvin (K), mole (mol), candela (cd), and ampere (A), used to measure length, time, mass, temperature, amount of a substance, light intensity, and electric current, respectively. This system was agreed upon in 1974 at an international general conference. There is another metric system, CGS, which stands for centimeter, gram, second; that system simply uses the hundredth of a meter (the centimeter) and the hundredth of the kilogram (the gram). The CGS system, formally introduced by the British Association for the Advancement of Science in 1874, is particularly useful to scientists making measurements of small quantities in laboratories, but it is less useful for space science. In this set, the SI system is used with the exception that temperatures will be presented in Celsius (C), instead of Kelvin. (The conversions between Celsius, Kelvin, and Fahrenheit temperatures are given below.) Often the standard unit of measure in the SI system, the meter, is too small when talking about the great distances in the solar system; kilometers (thousands of meters) or AU (astronomical units, defined below) will often be used instead of meters.

How is a unit defined? At one time a "meter" was defined as the length of a special metal ruler kept under strict conditions of temperature and humidity. That perfect meter could not be measured, however, without changing its temperature by opening the box, which would change its length, through thermal expansion or contraction. Today a meter is no longer defined according to a physical object; the only fundamental measurement that still is defined by a physical object is the kilogram. All of these units have had long and complex histories of attempts to define them. Some of the modern definitions, along with the use and abbreviation of each, are listed in the table here.

FUNDAMENTAL UNITS			
Measurement	Unit	Symbol	Definition
length	meter	m	The meter is the distance traveled by light in a vacuum during 1/299,792,458 of a second.
time	second	sec	The second is defined as the period of time in which the oscillations of cesium atoms, under specified conditions, complete exactly 9,192,631,770 cycles. The length of a second was thought to be a constant before Einstein developed theories in physics that show that the closer to the speed of light an object is traveling, the slower time is for that object. For the velocities on Earth, time is quite accurately still considered a constant.
mass	kilogram	kg	The International Bureau of Weights and Measures keeps the world's standard kilogram in Paris, and that object is the definition of the kilogram.
temperature	kelvin	K	A degree in Kelvin (and Celsius) is 1/273.16 of the thermody- namic temperature of the triple point of water (the temper- ature at which, under one atmosphere pressure, water coexists as water vapor, liquid, and solid ice). In 1967, the General Conference on Weights and Measures defined this temperature as 273.16 kelvin.

(continues)

FUNDAMENTAL UNITS (continued)

Measurement	Unit	Symbol	Definition
amount of a substance	mole	mol	The mole is the amount of a substance that contains as many units as there are atoms in 0.012 kilogram of carbon 12 (that is, Avogadro's number, or 6.02205×10^{23}). The units may be atoms, molecules, ions, or other particles.
electric current	ampere	A	The ampere is that constant current which, if maintained in two straight parallel conductors of infinite length, of negligible circular cross section, and placed one meter apart in a vacuum, would produce between these conductors a force equal to 2×10^{-7} newtons per meter of length.
light intensity	candela	cd	The candela is the luminous intensity of a source that emits monochromatic radiation with a wavelength of 555.17 nm and that has a radiant intensity of 1/683 watt per steradian. Normal human eyes are more sensitive to the yellow-green light of this wavelength than to any other.

Mass and weight are often confused. Weight is proportional to the force of gravity: Your weight on Earth is about six times your weight on the Moon because Earth's gravity is about six times that of the Moon's. Mass, on the other hand, is a quantity of matter, measured independently of gravity. In fact, weight has different units from mass: Weight is actually measured as a force (newtons, in SI, or pounds, in the English system).

The table "Fundamental Units" lists the fundamental units of the SI system. These are units that need to be defined in order to make other measurements. For example, the meter and the second are fundamental units (they are not based on any other units). To measure velocity, use a derived unit, meters per second (m/sec), a combination of fundamental units. Later in this section there is a list of common derived units.

The systems of temperature are capitalized (Fahrenheit, Celsius, and Kelvin), but the units are not (degree and kelvin). Unit abbreviations are capitalized only when they are named after a person, such as K for Lord Kelvin, or A for André-Marie Ampère. The units themselves are always lowercase, even when named for a person: one newton, or one N. Throughout these tables a small dot indicates multiplication, as in N m, which means a newton (N) times a meter (m). A space between the symbols can also be used to indicate multiplication, as in N • m. When a small letter is placed in front of a symbol, it is a prefix meaning some multiplication factor. For example, J stands for the unit of energy called a joule, and a mJ indicates a millijoule, or 10^{-3} joules. The table of prefixes is given at the end of this section.

COMPARISONS AMONG KELVIN, CELSIUS, AND FAHRENHEIT

One kelvin represents the same temperature difference as 1°C, and the temperature in kelvins is always equal to 273.15 plus the temperature in degrees Celsius. The Celsius scale was designed around the behavior of water. The freezing point of water (at one atmosphere of pressure) was originally defined to be 0°C, while the boiling point is 100°C. The kelvin equals exactly 1.8°F.

To convert temperatures in the Fahrenheit scale to the Celsius scale, use the following equation, where F is degrees Fahrenheit, and C is degrees Celsius:

$$C = (F - 32)/1.8 \ .$$

And to convert Celsius to Fahrenheit, use this equation:

$$F = 1.8C + 32 \ .$$

To convert temperatures in the Celsius scale to the Kelvin scale, add 273.16. By convention, the degree symbol (°) is used for Celsius and Fahrenheit temperatures but not for temperatures given in Kelvin, for example, 0°C equals 273K.

What exactly is temperature? Qualitatively, it is a measurement of how hot something feels, and this definition is so easy to relate to that people seldom take it further. What is really happening in a substance as it gets hot or cold, and how does that change make temperature? When a fixed amount of energy is put into a substance, it heats up by an amount depending on what it is. The temperature of an object, then, has something to do with how the material responds to energy, and that response is called entropy. The entropy of a material (entropy is usually denoted S) is a measure of atomic wiggling and disorder of the atoms in the material. Formally, temperature is defined as

$$\frac{1}{T} = \left(\frac{dS}{dU}\right)_N ,$$

meaning one over temperature (the reciprocal of temperature) is defined as the change in entropy (dS, in differential notation) per change in energy (dU), for a given number of atoms (N). What this means in less technical terms is that temperature is a measure of how much heat it takes to increase the entropy (atomic wiggling and disorder) of a substance. Some materials get hotter with less energy, and others require more to reach the same temperature.

The theoretical lower limit of temperature is -459.67°F (-273.15°C, or 0K), known also as absolute zero. This is the temperature at which all atomic movement stops. The Prussian physicist Walther Nernst showed that it is impossible to actually reach absolute zero, though with laboratory methods using nuclear magnetization it is possible to reach 10^{-6}K (0.000001K).

USEFUL MEASURES OF DISTANCE

A *kilometer* is a thousand meters (see the table "International System Prefixes"), and a *light-year* is the distance light travels in a vacuum during one year (exactly 299,792,458 m/sec, but commonly rounded to 300,000,000 m/sec). A light-year, therefore, is the distance that light can travel in one year, or:

*299,792,458 m/sec × 60 sec/min × 60 min/hr × 24 hr/
day × 365 days/yr = 9.4543 × 10¹⁵ m/yr.*

For shorter distances, some astronomers use light minutes and even light seconds. A light minute is 17,998,775 km, and a light second is 299,812.59 km. The nearest star to Earth, Proxima Centauri, is 4.2 light-years away from the Sun. The next, Rigil Centaurs, is 4.3 light-years away.

An *angstrom* (10^{-10}m) is a unit of length most commonly used in nuclear or particle physics. Its symbol is Å. The diameter of an atom is about one angstrom (though each element and isotope is slightly different).

An astronomical unit (AU) is a unit of distance used by astronomers to measure distances in the solar system. One astronomical unit equals the average distance from the center of the Earth to the center of the Sun. The currently accepted value, made standard in 1996, is 149,597,870,691 meters, plus or minus 30 meters.

One kilometer equals 0.62 miles, and one mile equals 1.61 kilometers.

The following table gives the most commonly used of the units derived from the fundamental units above (there are many more derived units not listed here because they have been developed for specific situations and are little-used elsewhere; for example, in the metric world, the curvature of a railroad track is measured with a unit called "degree of curvature," defined as the angle between two points in a curving track that are separated by a chord of 20 meters).

Though the units are given in alphabetical order for ease of reference, many can fit into one of several broad categories: dimensional units (angle, area, volume), material properties (density, viscosity, thermal expansivity), properties of motion (velocity, acceleration, angular velocity), electrical properties (frequency, electric charge, electric potential, resistance, inductance, electric field strength), magnetic properties (magnetic field strength, magnetic flux, magnetic flux density), and properties of radioactivity (amount of radioactivity and effect of radioactivity).

(continues on page 231)

DERIVED UNITS

Measurement	Unit symbol (derivation)	Comments
acceleration	unnamed (m/sec^2)	
angle	radian rad (m/m)	One radian is the angle centered in a circle that includes an arc of length equal to the radius. Since the circumference equals two pi times the radius, one radian equals 1/(2 pi) of the circle, or approximately 57.296°.
	steradian sr (m^2/m^2)	The steradian is a unit of solid angle. There are four pi steradians in a sphere. Thus one steradian equals about 0.079577 sphere, or about 3282.806 square degrees.
angular velocity	unnamed (rad/sec)	
area	unnamed (m^2)	
density	unnamed (kg/m^3)	Density is mass per volume. Lead is dense, styrofoam is not. Water has a density of one gram per cubic centimeter or 1,000 kilograms per cubic meter.
electric charge or electric flux	coulomb C ($A \cdot sec$)	One coulomb is the amount of charge accumulated in one second by a current of one ampere. One coulomb is also the amount of charge on 6.241506×10^{18} electrons.
electric field strength	unnamed [($kg \cdot m$)/($sec^3 \cdot A$) × V/m]	Electric field strength is a measure of the intensity of an electric field at a particular location. A field strength of one V/m represents a potential difference of one volt between points separated by one meter.
electric potential, or electromotive force (often called voltage)	volt V [($kg \cdot m^2$)/($sec^3 \cdot A$) = J/C = W/A]	Voltage is an expression of the potential difference in charge between two points in an electrical field. Electric potential is defined as the amount of potential energy present per unit of charge. One volt is a potential of one joule per coulomb of charge. The greater the voltage, the greater the flow of electrical current.

Measurement	Unit symbol (derivation)	Comments
energy, work, or heat	joule J [N·m (=kg·m²/sec²)]	
	electron volt eV	The electron volt, being so much smaller than the joule (one eV = 1.6 × 10⁻¹⁷ J), is useful for describing small systems.
force	newton N (kg·m/sec²)	This unit is the equivalent to the pound in the English system, since the pound is a measure of force and not mass.
frequency	hertz Hz (cycles/sec)	Frequency is related to wavelength as follows: kilohertz × wavelength in meters = 300,000.
inductance	henry H (Wb/A)	Inductance is the amount of magnetic flux a material pro- duces for a given current of electricity. Metal wire with an electric current passing through it creates a magnetic field; different types of metal make magnetic fields with different strengths and therefore have different inductances.
magnetic field strength	unnamed (A/m)	Magnetic field strength is the force that a magnetic field exerts on a theoretical unit magnetic pole.
magnetic flux	weber Wb [(kg·m²)/ (sec²·A) = V·sec]	The magnetic flux across a perpendicular surface is the product of the magnetic flux density, in teslas, and the surface area, in square meters.
magnetic flux density	tesla T [kg/(sec²·A) = Wb/m²]	A magnetic field of one tesla is strong: The strongest artificial fields made in laboratories are about 20 teslas, and the Earth's magnetic flux density, at its surface, is about 50 microteslas (μT). Planetary magnetic fields are sometimes measured in gammas, which are nanoteslas (10⁻⁹ teslas).
momentum, or impulse	unnamed [N·sec (= kg m/sec)]	Momentum is a measure of moving mass: how much mass and how fast it is moving.
power	watt W [J/s (= (kg m²)/sec³)]	Power is the rate at which energy is spent. Power can be mechanical (as in horsepower) or electrical (a watt is produced by a current of one ampere flowing through an electric potential of one volt).

	DERIVED UNITS *(continued)*	
Measurement	**Unit symbol (derivation)**	**Comments**
pressure, or stress	pascal Pa (N/m²)	The high pressures inside planets are often measured in gigapascals (10^9 pascals), abbreviated GPa. ~10,000 atm = one GPa.
	atmosphere atm	The atmosphere is a handy unit because one atmosphere is approximately the pressure felt from the air at sea level on Earth; one standard atm = 101,325 Pa; one metric atm = 98,066 Pa; one atm ~ one bar.
radiation per unit mass receiving it	gray (J/kg)	The amount of radiation energy absorbed per kilogram of mass. One gray = 100 rads, an older unit.
radiation (effect of)	sievert Sv	This unit is meant to make comparable the biological effects of different doses and types of radiation. It is the energy of radiation received per kilogram, in grays, multiplied by a factor that takes into consideration the damage done by the particular type of radiation.
radioactivity (amount)	becquerel Bq	One atomic decay per second
	curie Ci	The curie is the older unit of measure but is still frequently seen. One Ci = 3.7×10^{10} Bq.
resistance	ohm Ω (V/A)	Resistance is a material's unwillingness to pass electric current. Materials with high resistance become hot rather than allowing the current to pass and can make excellent heaters.
thermal expansivity	unnamed (/°)	This unit is per degree, measuring the change in volume of a substance with the rise in temperature.
vacuum	torr	Vacuum is atmospheric pressure below one atm (one torr = 1/760 atm). Given a pool of mercury with a glass tube standing in it, one torr of pressure on the pool will press the mercury one millimeter up into the tube, where one standard atmosphere will push up 760 millimeters of mercury.

Measurement	Unit symbol (derivation)	Comments
velocity	unnamed (m/sec)	
viscosity	unnamed [Pa·sec (= kg/ (m·sec)]	Viscosity is a measure of resistance to flow. If a force of one newton is needed to move one square meter of the liquid or gas relative to a second layer one meter away at a speed of one meter per second, then its viscosity is one Pa·s, often simply written Pas or Pas. The cgs unit for viscosity is the poise, equal to 0.1Pa·s.
volume	cubic meter (m³)	

(continued from page 227)

DEFINITIONS FOR ELECTRICITY AND MAGNETISM

When two objects in each other's vicinity have different electrical charges, an *electric field* exists between them. An electric field also forms around any single object that is electrically charged with respect to its environment. An object is negatively charged (-) if it has an excess of electrons relative to its surroundings. An object is positively charged (+) if it is deficient in electrons with respect to its surroundings.

An electric field has an effect on other charged objects in the vicinity. The field strength at a particular distance from an object is directly proportional to the electric charge of that object, in coulombs. The field strength is inversely proportional to the distance from a charged object.

Flux is the rate (per unit of time) in which something flowing crosses a surface perpendicular to the direction of flow.

An alternative expression for the intensity of an electric field is *electric flux density*. This refers to the number of lines of electric flux passing at right angles through a given surface area, usually one meter squared (1 m²). Electric flux density, like electric field strength, is directly proportional to the charge on the object. But flux density diminishes with distance according to the inverse-square law because it is specified in

INTERNATIONAL SYSTEM PREFIXES

SI prefix	Symbol	Multiplying factor
exa-	E	$10^{18} = 1{,}000{,}000{,}000{,}000{,}000{,}000$
peta-	P	$10^{15} = 1{,}000{,}000{,}000{,}000{,}000$
tera-	T	$10^{12} = 1{,}000{,}000{,}000{,}000$
giga-	G	$10^{9} = 1{,}000{,}000{,}000$
mega-	M	$10^{6} = 1{,}000{,}000$
kilo-	k	$10^{3} = 1{,}000$
hecto-	h	$10^{2} = 100$
deca-	da	$10 = 10$
deci-	d	$10^{-1} = 0.1$
centi-	c	$10^{-2} = 0.01$
milli-	m	$10^{-3} = 0.001$
micro-	μ or u	$10^{-6} = 0.000{,}001$
nano-	n	$10^{-9} = 0.000{,}000{,}001$
pico-	p	$10^{-12} = 0.000{,}000{,}000{,}001$
femto-	f	$10^{-15} = 0.000{,}000{,}000{,}000{,}001$
atto-	a	$10^{-18} = 0.000{,}000{,}000{,}000{,}000{,}001$

A note on nonmetric prefixes: In the United States, the word billion means the number 1,000,000,000, or 10^9. In most countries of Europe and Latin America, this number is called "one milliard" or "one thousand million," and "billion" means the number 1,000,000,000,000, or 10^{12}, which is what Americans call a "trillion." In this set, a billion is 10^9.

terms of a surface area (per meter squared) rather than a linear displacement (per meter).

A *magnetic field* is generated when electric charge carriers such as electrons move through space or within an electrical conductor. The geometric shapes of the magnetic flux lines produced by moving charge carriers (electric current) are similar to the shapes of the flux lines in an electrostatic field. But

NAMES FOR LARGE NUMBERS

Number	American	European	SI prefix
10^9	billion	milliard	giga-
10^{12}	trillion	billion	tera-
10^{15}	quadrillion	billiard	peta-
10^{18}	quintillion	trillion	exa-
10^{21}	sextillion	trilliard	zetta-
10^{24}	septillion	quadrillion	yotta-
10^{27}	octillion	quadrilliard	
10^{30}	nonillion	quintillion	
10^{33}	decillion	quintilliard	
10^{36}	undecillion	sextillion	
10^{39}	duodecillion	sextilliard	
10^{42}	tredecillion	septillion	
10^{45}	quattuordecillion	septilliard	

This naming system is designed to expand indefinitely by factors of powers of three. Then, there is also the googol, the number 10^{100} (one followed by 100 zeroes). The googol was invented for fun by the eight-year-old nephew of the American mathematician Edward Kasner. The googolplex is 10^{googol}, or one followed by a googol of zeroes. Both it and the googol are numbers larger than the total number of atoms in the universe, thought to be about 10^{80}.

there are differences in the ways electrostatic and magnetic fields interact with the environment.

Electrostatic flux is impeded or blocked by metallic objects. *Magnetic flux* passes through most metals with little or no effect, with certain exceptions, notably iron and nickel. These two metals, and alloys and mixtures containing them, are known as ferromagnetic materials because they concentrate magnetic lines of flux.

Magnetic flux density and *magnetic force* are related to *magnetic field strength*. In general, the magnetic field strength diminishes with increasing distance from the axis of a magnetic dipole in which the flux field is stable. The function defining the rate at which this field-strength decrease occurs depends on the geometry of the magnetic lines of flux (the shape of the flux field).

PREFIXES

Adding a prefix to the name of that unit forms a multiple of a unit in the International System (see the table "International System Prefixes"). The prefixes change the magnitude of the unit by orders of 10 from 10^{18} to 10^{-18}.

Very small concentrations of chemicals are also measured in parts per million (ppm) or parts per billion (ppb), which mean just what they sound like: If there are four parts per million of lead in a rock (4 ppm), then out of every million atoms in that rock, on average four of them will be lead.

APPENDIX 2:

Light, Wavelength, and Radiation

Electromagnetic radiation is energy given off by matter, traveling in the form of waves or particles. Electromagnetic energy exists in a wide range of energy values, of which visible light is one small part of the total spectrum. The source of radiation may be the hot and therefore highly energized atoms of the Sun, pouring out radiation across a wide range of energy values, including of course visible light, and they may also be unstable (radioactive) elements giving off radiation as they decay.

Radiation is called "electromagnetic" because it moves as interlocked waves of electrical and magnetic fields. A wave is a disturbance traveling through space, transferring energy from one point to the next. In a vacuum, all electromagnetic radiation travels at the speed of light, 983,319,262 feet per second (299,792,458 m/sec, often approximated as 300,000,000 m/sec). Depending on the type of radiation, the waves have different wavelengths, energies, and frequencies (see the following figure). The wavelength is the distance between individual waves, from one peak to another. The frequency is the number of waves that pass a stationary point each second. Notice in the graphic how the wave undulates up and down from peaks to valleys to peaks. The time from one peak to the next peak is called one cycle. A single unit of frequency is equal to one cycle per second. Scientists refer to a single cycle as one hertz, which commemorates 19th-century German physicist Heinrich Hertz, whose discovery of electromagnetic waves led to the development of radio. The frequency of a wave is related to its energy: The higher the frequency of a

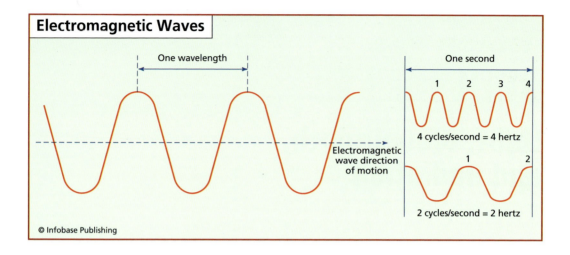

Electromagnetic Waves

One wavelength

One second

1 2 3 4

4 cycles/second = 4 hertz

1 2

2 cycles/second = 2 hertz

Electromagnetic
wave direction
of motion

© Infobase Publishing

Each electromagnetic wave has a measurable wavelength and frequency.

wave, the higher its energy, though its speed in a vacuum does not change.

The smallest wavelength, highest energy and frequency electromagnetic waves are cosmic rays, then as wavelength increases and energy and frequency decrease, come gamma rays, then X-rays, then ultraviolet light, then visible light (moving from violet through indigo, blue, green, yellow, orange, and red), then infrared (divided into near, meaning near to visible, mid-, and far infrared), then microwaves, and then radio waves, which have the longest wavelengths and the lowest energy and frequency. The electromagnetic spectrum is shown in the accompanying figure and table.

As a wave travels and vibrates up and down with its characteristic wavelength, it can be imagined as vibrating up and down in a single plane, such as the plane of this sheet of paper in the case of the simple example in the figure here showing polarization. In nature, some waves change their polarization constantly so that their polarization sweeps through all angles, and they are said to be circularly polarized. In ordinary visible light, the waves are vibrating up and down in numerous random planes. Light can be shone through a special filter called a polarizing filter that blocks out all the light except that polarized in a certain direction, and the light that shines out the other side of the filter is then called polarized light.

Polarization is important in wireless communications systems such as radios, cell phones, and non-cable television. The orientation of the transmitting antenna creates the polarization of the radio waves transmitted by that antenna: A vertical antenna emits vertically polarized waves, and a horizontal antenna emits horizontally polarized waves. Similarly, a horizontal antenna is best at receiving horizontally polarized waves and a vertical antenna at vertically polarized waves. The best communications are obtained when the source and receiver antennas have the same polarization. This is why,

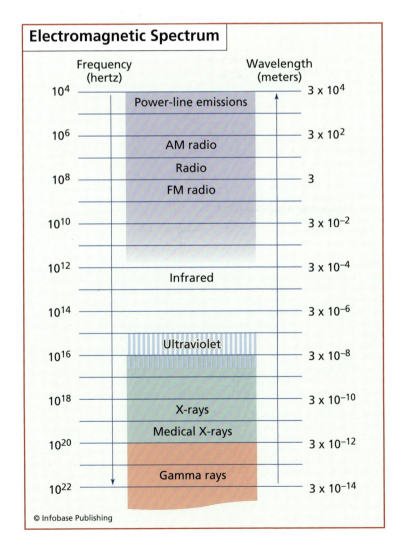

Electromagnetic Spectrum

Frequency (hertz)		Wavelength (meters)
10^4	Power-line emissions	3×10^4
10^6	AM radio	3×10^2
	Radio	
10^8	FM radio	3
10^{10}		3×10^{-2}
10^{12}	Infrared	3×10^{-4}
10^{14}		3×10^{-6}
10^{16}	Ultraviolet	3×10^{-8}
10^{18}	X-rays	3×10^{-10}
10^{20}	Medical X-rays	3×10^{-12}
10^{22}	Gamma rays	3×10^{-14}

© Infobase Publishing

Electromagnetic spectrum ranges from cosmic rays at the shortest wavelengths to radiowaves at the longest wavelengths.

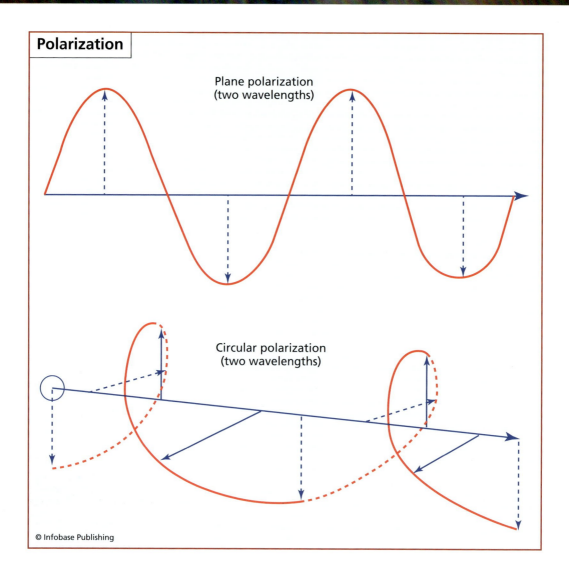

Polarization

Plane polarization
(two wavelengths)

Circular polarization
(two wavelengths)

© Infobase Publishing

Waves can be thought of as plane or circularly polarized.

when trying to adjust television antennas to get a better signal, having the two antennae at right angles to each other can maximize the chances of receiving a signal.

The human eye stops being able to detect radiation at wavelengths between 3,000 and 4,000 angstroms, which is deep violet—also the rough limit on transmissions through the atmosphere. (See the table "Wavelengths and Frequencies of Visible Light.") (Three thousand to 4,000 angstroms is the same as 300–400 nm because an angstrom is 10^{-9} m, while the prefix nano- or n means 10^{-10}; for more, see appendix 1, "Units and Measurements.") Of visible light, the colors red, orange, yellow, green, blue, indigo, and violet are listed in order from longest wavelength and lowest energy to shortest wavelength and highest energy. Sir Isaac Newton, the spectacular English physicist and mathematician, first found that a glass prism split sunlight into a rainbow of colors. He named this a "spectrum," after the Latin word for ghost.

If visible light strikes molecules of gas as it passes through the atmosphere, it may get absorbed as energy by the molecule. After a short amount of time, the molecule releases the light, most probably in a different direction. The color that is radiated is the same color that was absorbed. All the colors of visible light can be absorbed by atmospheric molecules, but the higher energy blue light is absorbed more often than the lower energy red light. This process is called Rayleigh scattering (named after Lord John Rayleigh, an English physicist who first described it in the 1870s).

The blue color of the sky is due to Rayleigh scattering. As light moves through the atmosphere, most of the longer wavelengths pass straight through: The air affects little of the red, orange, and yellow light. The gas molecules absorb

WAVELENGTHS AND FREQUENCIES OF VISIBLE LIGHT

Visible light color	Wavelength (in Å, angstroms)	Frequency (times 10^{14} Hz)
violet	4,000–4,600	7.5–6.5
indigo	4,600–4,750	6.5–6.3
blue	4,750–4,900	6.3–6.1
green	4,900–5,650	6.1–5.3
yellow	5,650–5,750	5.3–5.2
orange	5,750–6,000	5.2–5.0
red	6,000–8,000	5.0–3.7

WAVELENGTHS AND FREQUENCIES OF THE ELECTROMAGNETIC SPECTRUM

Energy	Frequency in hertz (Hz)	Wavelength in meters
cosmic rays	everything higher in energy than gamma rays	everything lower in wavelength than gamma rays
gamma rays	10^{20} to 10^{24}	less than 10^{-12} m
X-rays	10^{17} to 10^{20}	1 nm to 1 pm
ultraviolet	10^{15} to 10^{17}	400 nm to 1 nm
visible	4×10^{14} to 7.5×10^{14}	750 nm to 400 nm
near-infrared	1×10^{14} to 4×10^{14}	2.5 µm to 750 nm
infrared	10^{13} to 10^{14}	25 µm to 2.5 µm
microwaves	3×10^{11} to 10^{13}	1 mm to 25 µm
radio waves	less than 3×10^{11}	more than 1 mm

much of the shorter wavelength blue light. The absorbed blue light is then radiated in different directions and is scattered all around the sky. Whichever direction you look, some of this scattered blue light reaches you. Since you see the blue light from everywhere overhead, the sky looks blue. Note also that there is a very different kind of scattering, in which the light is simply bounced off larger objects like pieces of dust and water droplets, rather than being absorbed by a molecule of gas in the atmosphere and then reemitted. This bouncing kind of scattering is responsible for red sunrises and sunsets.

Until the end of the 18th century, people thought that visible light was the only kind of light. The amazing amateur astronomer Frederick William Herschel (the discoverer of Uranus) discovered the first non-visible light, the infrared. He thought that each color of visible light had a different temperature and devised an experiment to measure the temperature of each color of light. The temperatures went up as the colors progressed from violet through red, and then Herschel decided to measure past red, where he found the highest temperature yet. This was the first demonstration that there was a kind of radiation that could not be seen by the human eye. Herschel originally named this range of radiation "calorific rays," but the name was later changed to infrared, meaning "below red."

COMMON USES FOR RADIO WAVES

User	Approximate frequency
AM radio	0.535×10^6 to 1.7×10^6 Hz
baby monitors	49×10^6 Hz
cordless phones	49×10^6 Hz 900×10^6 Hz $2,400 \times 10^6$ Hz
television channels 2 through 6	54×10^6 to 88×10^6 Hz
radio-controlled planes	72×10^6 Hz
radio-controlled cars	75×10^6 Hz
FM radio	88×10^6 to 108×10^6 Hz
television channels 7 through 13	174×10^6 to 220×10^6 Hz
wildlife tracking collars	215×10^6 Hz
cell phones	800×10^6 Hz $2,400 \times 10^6$ Hz
air traffic control radar	960×10^6 Hz $1,215 \times 10^6$ Hz
global positioning systems	$1,227 \times 10^6$ Hz $1,575 \times 10^6$ Hz
deep space radio	$2,300 \times 10^6$ Hz

Infrared radiation has become an important way of sensing solar system objects and is also used in night-vision goggles and various other practical purposes.

At lower energies and longer wavelengths than the visible and infrared, microwaves are commonly used to transmit energy to food in microwave ovens, as well as for some communications, though radio waves are more common in this use. There is a wide range of frequencies in the radio spectrum, and they are used in many ways, as shown in the table "Common Uses for Radio Waves," including television, radio, and cell phone transmissions. Note that the frequency units are given in terms of 10^6 Hz, without correcting for each coefficient's additional factors of 10. This is because 10^6 Hz corresponds to the unit of megahertz (MHz), which is a commonly used unit of frequency.

Cosmic rays, gamma rays, and X-rays, the three highest-energy radiations, are known as ionizing radiation because they contain enough energy that, when they hit an atom, they may knock an electron off of it or otherwise change the atom's weight or structure. These ionizing radiations, then, are particularly dangerous to living things; for example, they can damage DNA molecules (though good use is made of them as well, to see into bodies with X-rays and to kill cancer cells with gamma rays). Luckily the atmosphere stops most ionizing radiation, but not all of it. Cosmic rays created by the Sun in solar flares, or sent off as a part of the solar wind, are relatively low energy. There are far more energetic cosmic rays, though, that come from distant stars through interstellar space. These are energetic enough to penetrate into an asteroid as deeply as a meter and can often make it through the atmosphere.

When an atom of a radioisotope decays, it gives off some of its excess energy as radiation in the form of X-rays, gamma rays, or fast-moving subatomic particles: alpha particles (two protons and two neutrons, bound together as an atomic *nucleus*), or beta particles (fast-moving electrons), or a combination of two or more of these products. If it decays with emission of an alpha or beta particle, it becomes a new element. These decay products can be described as gamma, beta, and alpha radiation. By decaying, the atom is progressing in

RADIOACTIVITY OF SELECTED OBJECTS AND MATERIALS

Object or material	Radioactivity
1 adult human (100 Bq/kg)	7,000 Bq
1 kg coffee	1,000 Bq
1 kg high-phosphate fertilizer	5,000 Bq
1 household smoke detector (with the element americium)	30,000 Bq
radioisotope source for cancer therapy	100 million million Bq
1 kg 50-year-old vitrified high-level nuclear waste	10 million million Bq
1 kg uranium ore (Canadian ore, 15% uranium)	25 million Bq
1 kg uranium ore (Australian ore, 0.3% uranium)	500,000 Bq
1 kg granite	1,000 Bq

one or more steps toward a stable state where it is no longer radioactive.

The X-rays and gamma rays from decaying atoms are identical to those from other natural sources. Like other ionizing radiation, they can damage living tissue but can be blocked by lead sheets or by thick concrete. Alpha particles are much larger and can be blocked more quickly by other material; a sheet of paper or the outer layer of skin on your hand will stop them. If the atom that produces them is taken inside the body, however, such as when a person breathes in radon gas, the alpha particle can do damage to the lungs. Beta particles are more energetic and smaller and can penetrate a couple of centimeters into a person's body.

But why can both radioactive decay that is formed of subatomic particles and heat that travels as a wave of energy be considered radiation? One of Albert Einstein's great discoveries is called the photoelectric effect: Subatomic particles can all behave as either a wave or a particle. The smaller the particle, the more wavelike it is. The best example of this is light itself, which behaves almost entirely as a wave, but there is the particle equivalent for light, the massless photon. Even alpha particles, the largest decay product discussed here, can act like a wave, though their wavelike properties are much harder to detect.

The amount of radioactive material is given in becquerel (Bq), a measure that enables us to compare the typical radioactivity of some natural and other materials. A becquerel is one atomic decay per second. Radioactivity is still sometimes measured using a unit called a Curie; a Becquerel is 27×10^{-12} Curies. There are materials made mainly of radioactive elements, like uranium, but most materials are made mainly of stable atoms. Even materials made mainly of stable atoms, however, almost always have trace amounts of radioactive elements in them, and so even common objects give off some level of radiation, as shown in the following table.

Background radiation is all around us all the time. Naturally occurring radioactive elements are more common in some kinds of rocks than others; for example, *granite* carries more radioactive elements than does sandstone; therefore a person working in a bank built of granite will receive more radiation than someone who works in a wooden building. Similarly, the atmosphere absorbs cosmic rays, but the higher the elevation, the more cosmic-ray exposure there is. A person living in Denver or in the mountains of Tibet is exposed to more cosmic rays than someone living in Boston or in the Netherlands.

APPENDIX 3:

A List of All Known Moons

Though Mercury and Venus have no moons, the other planets in the solar system have at least one. Some moons, such as Earth's Moon and Jupiter's Galileans satellites, are thought to have formed at the same time as their accompanying planet. Many other moons appear simply to be captured asteroids; for at least half of Jupiter's moons, this seems to be the case. These small, irregular moons are difficult to detect from Earth, and so the lists given in the table below must be considered works in progress for the gas giant planets. More moons will certainly be discovered with longer observation and better instrumentation.

KNOWN MOONS OF ALL PLANETS						
Earth	Mars	Jupiter	Saturn	Uranus	Neptune	Pluto
1	2	63	62	27	13	3
1. Moon	1. Phobos	1. Metis	1. S/2009 S1	1. Cordelia	1. Naiad	1. Charon
	2. Diemos	2. Adrastea	2. Pan	2. Ophelia	2. Thalassa	2. Nix (P1)
		3. Amalthea	3. Daphnis	3. Bianca	3. Despina	3. Hydra
		4. Thebe	4. Atlas	4. Cressida	4. Galatea	(P2)
		5. Io	5. Prometheus	5. Desdemona	5. Larissa	
		6. Europa	6. Pandora	6. Juliet	6. Proteus	
		7. Ganymede	7. Epimetheus	7. Portia	7. Triton	
		8. Callisto	8. Janus	8. Rosalind	8. Nereid	
		9. Themisto	9. Aegaeon	9. Cupid (2003 U2)	9. Halimede (S/2002 N1)	
		10. Leda	10. Mimas	10. Belinda		
		11. Himalia	11. Methone			

(continues)

KNOWN MOONS OF ALL PLANETS *(continued)*

Earth	Mars	Jupiter	Saturn	Uranus	Neptune	Pluto
		12. Lysithea	12. Anthe	11. Perdita (1986 U10)	10. Sao (S/2002 N2)	
		13. Elara	13. Pallene	12. Puck	11. Laomedeia (S/2002 N3)	
		14. S/2000 J11	14. Enceladus	13. Mab (2003 U1)	12. Psamathe (S/2003 N1)	
		15. Carpo (S/2003 J20)	15. Telesto	14. Miranda	13. Neso (S/2002 N4)	
		16. S/2003 J12	16. Tethys	15. Ariel		
		17. Euporie	17. Calypso	16. Umbriel		
		18. S/2003 J3	18. Dione	17. Titania		
		19. S/2003 J18	19. Helene	18. Oberon		
		20. Orthosie	20. Polydeuces	19. Francisco (2001 U3)		
		21. Euanthe	21. Rhea	20. Caliban		
		22. Harpalyke	22. Titan	21. Stephano		
		23. Praxidike	23. Hyperion	22. Trinculo		
		24. Thyone	24. Iapetus	23. Sycorax		
		25. S/2003 J16	25. Kiviuq	24. Margaret (2003 U3)		
		26. Mneme (S/2003 J21)	26. Ijiraq	25. Prospero		
		27. Iocaste	27. Phoebe	26. Setebos		
		28. Helike (S/2003 J6)	28. Paaliaq	27. Ferdinand (2001 U2)		
		29. Hermippe	29. Skathi			
		30. Thelxinoe (S/2003 J22)	30. Albiorix			
		31. Ananke	31. S/2007 S2			
		32. S/2003 J15	32. Bebhionn			
		33. Eurydome	33. Erriapo			
		34. S/2003 J17	34. Siarnaq			
		35. Pasithee	35. Skoll			
		36. S/2003 J10	36. Tarvos			
			37. Tarqeq			
			38. Greip			
			39. Hyrrokkin			
			40. S/2004 S13			
			41. S/2004 S17			
			42. Mundilfari			
			43. Jarnsaxa			
			44. S/2006 S1			

Earth	Mars	Jupiter	Saturn	Uranus	Neptune	Pluto
		37. Chaldene	45. Narvi			
		38. Isonoe	46. Bergelmir			
		39. Erinome	47. Suttungr			
		40. Kale	48. S/2004			
		41. Aitne	S12			
		42. Taygete	49. S/2004			
		43. Kallichore	S7			
		(S/2003	50. Hati			
		J11)	51. Bestla			
		44. Eukelade	52. Farbauti			
		(S/2003 J1)	53. Thrymyr			
		45. Arche	54. S/2007			
		(S/2002 J1)	S3			
		46. S/2003 J9	55. Aegir			
		47. Carme	56. S/2006			
		48. Kalyke	S3			
		49. Sponde	57. Kari			
		50. Magaclite	58. Fenrir			
		51. S/2003 J5	59. Surtur			
		52. S/2003	60. Ymir			
		J19	61. Loge			
		53. S/2003	62. Fornjot			
		J23				
		54. Hegemone				
		(S/2003 J8)				
		55. Pasiphae				
		56. Cyllene				
		(S/2003				
		J13)				
		57. S/2003 J4				
		58. Sinope				
		59. Aoede				
		(S/2003 J7)				
		60. Autonoe				
		61. Calirrhoe				
		62. Kore				
		(S/2003				
		J14)				
		63. S/2003 J2				

Glossary

accretion The accumulation of celestial gas, dust, or smaller bodies by gravitational attraction into a larger body, such as a planet or an asteroid

achondite A stony (silicate-based) meteorite that contains no chondrules; these originate in differentiated bodies and may be mantle material or lavas (see also CHONDRITE and IRON METEORITE)

albedo The light reflected by an object as a fraction of the light shining on an object; mirrors have high albedo, while charcoal has low albedo

anorthite A calcium-rich plagioclase mineral with compositional formula $CaAl_2Si_2O_8$, significant for making up the majority of the rock anorthosite in the crust of the Moon

anticyclone An area of increased atmospheric pressure relative to the surrounding pressure field in the atmosphere, resulting in circular flow in a clockwise direction north of the equator and in a counterclockwise direction to the south

aphelion A distance; the farthest from the Sun an object travels in its orbit

apogee As for aphelion but for any orbital system (not confined to the Sun)

apparent magnitude The brightness of a celestial object as it would appear from a given distance—the lower the number, the brighter the object

atom The smallest quantity of an element that can take part in a chemical reaction; consists of a nucleus of protons and neutrons, surrounded by a cloud of electrons; each atom is about 10^{-10} meters in diameter, or one angstrom

atomic number   The number of protons in an atom's nucleus

AU   An AU is an astronomical unit, defined as the distance from the Sun to the Earth; approximately 93 million miles, or 150 million kilometers. For more information, refer to the UNITS AND MEASUREMENTS appendix

basalt   A generally dark-colored extrusive igneous rock most commonly created by melting a planet's mantle; its low silica content indicates that it has not been significantly altered on its passage to the planet's surface

bolide   An object falling into a planet's atmosphere, when a specific identification as a comet or asteroid cannot be made

bow shock   The area of compression in a flowing fluid when it strikes an object or another fluid flowing at another rate; for example, the bow of a boat and the water, or the magnetic field of a planet and the flowing solar wind

breccia   Material that has been shattered from grinding, as in a fault, or from impact, as by meteorites or other solar system bodies

CAIs (calcium-aluminum inclusions)   Small spheres of mineral grains found in chondritic meteorites and thought to be the first solids that formed in the protoplanetary disk

calcium-aluminum inclusion   See CAIs.

chondrite   A class of meteorite thought to contain the most primitive material left from the solar nebula; named after their glassy, super-primitive inclusions called chondrules

chondrule   Rounded, glassy, and crystalline bodies incorporated into the more primitive of meteorites; thought to be the condensed droplets of the earliest solar system materials

CI chondrite   The class of chondrite meteorites with compositions most like the Sun and therefore thought to be the oldest and least altered material in the solar system

clinopyroxene   A common mineral in the mantle and igneous rocks, with compositional formula $((Ca,Mg,Fe,Al)_2(Si,Al)_2O_6)$

conjunction   When the Sun is between the Earth and the planet or another body in question

convection Material circulation upward and downward in a gravity field caused by horizontal gradients in density; an example is the hot, less dense bubbles that form at the bottom of a pot, rise, and are replaced by cooler, denser sinking material

core The innermost material within a differentiated body; in a rocky planet this consists of iron-nickel metal, and in a gas planet this consists of the rocky innermost solids

Coriolis force The effect of movement on a rotating sphere; movement in the Northern Hemisphere curves to the right, while movement in the Southern Hemisphere curves to the left

craton The ancient, stable interior cores of the Earth's continents

crust The outermost layer of most differentiated bodies, often consisting of the least dense products of volcanic events or other buoyant material

cryovolcanism Non-silicate materials erupted from icy and gassy bodies in the cold outer solar system; for example, as suspected or seen on the moons Enceladus, Europa, Titan, and Triton

cubewano Any large Kuiper belt object orbiting between about 41 AU and 48 AU but not controlled by orbital resonances with Neptune; the odd name is derived from 1992 QB_1, the first Kuiper belt object found

cyclone An area in the atmosphere in which the pressures are lower than those of the surrounding region at the same level, resulting in circular motion in a counterclockwise direction north of the equator and in a clockwise direction to the south

debris disk A flattened, spinning disk of dust and gas around a star formed from collisions among bodies already accreted in an aging solar system

differential rotation Rotation at different rates at different latitudes, requiring a liquid or gassy body, such as the Sun or Jupiter

differentiated body   A spherical body that has a structure of concentric spherical layers, differing in terms of composition, heat, density, and/or motion; caused by gravitational separations and heating events such as planetary accretion

dipole   Two associated magnetic poles, one positive and one negative, creating a magnetic field

direct (prograde)   Rotation or orbit in the same direction as the Earth's, that is, counterclockwise when viewed from above its North Pole

disk wind   Magnetic fields that either pull material into the protostar or push it into the outer disk; these are thought to form at the inner edge of the disk where the protostar's magnetic field crosses the disk's magnetic field (also called X-WIND)

distributary   River channels that branch from the main river channel, carrying flow away from the central channel; usually form fans of channels at a river's delta

eccentricity   The amount by which an ellipse differs from a circle

ecliptic   The imaginary plane that contains the Earth's orbit and from which the planes of other planets' orbits deviate slightly; the ecliptic makes an angle of seven degrees with the plane of the Sun's equator

ejecta   Material thrown out of the site of a crater by the force of the impactor

element   A family of atoms that all have the same number of positively charged particles in their nuclei (the center of the atom)

ellipticity   The amount by which a planet's shape deviates from a sphere

equinox   One of two points in a planet's orbit when day and night have the same length; vernal equinox occurs in Earth's spring and autumnal equinox in the fall

exosphere   The uppermost layer of a planet's atmosphere

extrasolar   Outside this solar system

faint young Sun paradox The apparent contradiction between the observation that the Sun gave off far less heat in its early years and the likelihood that the Earth was still warm enough to host liquid water

garnet The red, green, or purple mineral that contains the majority of the aluminum in the Earth's upper mantle; its compositional formula is $((Ca,Mg,Fe\ Mn)_3(Al,Fe,Cr,Ti)_2(SiO_4)3)$

giant molecular cloud An interstellar cloud of dust and gas that is the birthplace of clusters of new stars as it collapses through its own gravity

graben A low area longer than it is wide and bounded from adjoining higher areas by faults; caused by extension in the crust

granite An intrusive igneous rock with high silica content and some minerals containing water; in this solar system thought to be found only on Earth

half-life The time it takes for half a population of an unstable isotope to decay

hydrogen burning The most basic process of nuclear fusion in the cores of stars that produces helium and radiation from hydrogen

igneous rock Rock that was once hot enough to be completely molten

impactor A generic term for an object striking and creating a crater in another body

inclination As commonly used in planetary science, the angle between the plane of a planet's orbit and the plane of the ecliptic

ionosphere The uppermost atmosphere of a planet where most gases exist as ionized particles and electrons; on Earth the ionosphere begins at about 50 miles (80 km) altitude

iron meteorite Meteorites that consist largely of iron-nickel metal; thought to be parts of the cores of smashed planetesimals from early solar system accretion

isotope Atoms with the same number of protons (therefore the same type of element) but different numbers of neutrons;

may be stable or radioactive and occur in different relative abundances

lander A spacecraft designed to land on another solar system object rather than flying by, orbiting, or entering the atmosphere and then burning up or crashing

lithosphere The uppermost layer of a terrestrial planet consisting of stiff material that moves as one unit if there are plate tectonic forces and does not convect internally but transfers heat from the planet's interior through conduction

magnetic moment The torque (turning force) exerted on a magnet when it is placed in a magnetic field

magnetopause The surface between the magnetosheath and the magnetosphere of a planet

magnetosheath The compressed, heated portion of the solar wind where it piles up against a planetary magnetic field

magnetosphere The volume of a planet's magnetic field, shaped by the internal planetary source of the magnetism and by interactions with the solar wind

magnitude See APPARENT MAGNITUDE

mantle The spherical shell of a terrestrial planet between crust and core; thought to consist mainly of silicate minerals

mass number The number of protons plus neutrons in an atom's nucleus

mesosphere The atmospheric layer between the stratosphere and the thermosphere

metal 1) Material with high electrical conductivity in which the atomic nuclei are surrounded by a cloud of electrons, that is, metallic bonds, or 2) in astronomy, any element heavier than helium

metallicity The fraction of all elements heavier than hydrogen and helium in a star or protoplanetary disk; higher metallicity is thought to encourage the formation of planets

metamorphic rock Rock that has been changed from its original state by heat or pressure but was never liquid

mid-ocean ridge The line of active volcanism in oceanic basins from which two oceanic plates are produced, one moving away from each side of the ridge; only exist on Earth

mineral   A naturally occurring inorganic substance having an orderly internal structure (usually crystalline) and characteristic chemical composition

nucleus   The center of the atom, consisting of protons (positively charged) and neutrons (no electric charge); tiny in volume but makes up almost all the mass of the atom

nutation   The slow wobble of a planet's rotation axis along a line of longitude, causing changes in the planet's obliquity

obliquity   The angle between a planet's equatorial plane to its orbit plane

occultation   The movement of one celestial body in front of another from a particular point of view; most commonly the movement of a planet in front of a star from the point of view of an Earth viewer

olivine   Also known as the gem peridot, the green mineral that makes up the majority of the upper mantle; its compositional formula is $((Mg, Fe)_2SiO_4)$

one-plate planet   A planet with lithosphere that forms a continuous spherical shell around the whole planet, not breaking into plates or moving with tectonics; Mercury, Venus, and Mars are examples

opposition   When the Earth is between the Sun and the planet of interest

orbital period   The time required for an object to make a complete circuit along its orbit

pallasite   A type of iron meteorite that also contains the silicate mineral olivine, and is thought to be part of the region between the mantle and core in a differentiated planetesimal that was shattered in the early years of the solar system

parent body   The larger body that has been broken to produce smaller pieces; large bodies in the asteroid belt are thought to be the parent bodies of meteorites that fall to Earth today

perigee   As for perihelion but for any orbital system (not confined to the Sun)

perihelion (pl. perihelia)   A distance; the closest approach to the Sun made in an object's orbit

planetary nebula   A shell of gas ejected from stars at the end of their lifetimes; unfortunately named in an era of primitive telescopes that could not discern the size and nature of these objects

planetesimal   The small, condensed bodies that formed early in the solar system and presumably accreted to make the planets; probably resembled comets or asteroids

plate tectonics   The movement of lithospheric plates relative to each other, only known on Earth

precession   The movement of a planet's axis of rotation that causes the axis to change its direction of tilt, much as the direction of the axis of a toy top rotates as it slows

primordial disk   Another name for a protoplanetary disk

prograde (direct)   Rotates or orbits in the same direction the Earth does, that is, counterclockwise when viewed from above its North Pole

proplyd   Abbreviation for a *protoplanetary disk*

protoplanetary disk   The flattened, spinning cloud of dust and gas surrounding a growing new star

protostar   The central mass of gas and dust in a newly forming solar system that will eventually begin thermonuclear fusion and become a star

radioactive   An atom prone to radiodecay

radio-decay   The conversion of an atom into a different atom or isotope through emission of energy or subatomic particles

red, reddened   A solar system body with a redder color in visible light, but more important, one that has increased albedo at low wavelengths (the "red" end of the spectrum)

reflectance spectra   The spectrum of radiation that bounces off a surface, for example, sunlight bouncing off the surface of an asteroid; the wavelengths with low intensities show the kinds of radiation absorbed rather than reflected by the surface and indicate the composition of the surface materials

refractory An element that requires unusually high temperatures in order to melt or evaporate; compare to volatile

relief (topographic relief) The shapes of the surface of land; most especially the high parts such as hills or mountains

resonance When the ratio of the orbital periods of two bodies is an integer; for example, if one moon orbits its planet once for every two times another moon orbits, the two are said to be in resonance

retrograde Rotates or orbits in the opposite direction to Earth, that is, clockwise when viewed from above its North Pole

Roche limit The radius around a given planet that a given satellite must be outside of in order to remain intact; within the Roche limit, the satellite's self-gravity will be overcome by gravitational tidal forces from the planet, and the satellite will be torn apart

rock Material consisting of the aggregate of minerals

sedimentary rock Rock made of mineral grains that were transported by water or air

seismic waves Waves of energy propagating through a planet, caused by earthquakes or other impulsive forces, such as meteorite impacts and human-made explosions

semimajor axis Half the widest diameter of an orbit

semiminor axis Half the narrowest diameter of an orbit

silicate A molecule, crystal, or compound made from the basic building block silica (SiO_2); the Earth's mantle is made of silicates, while its core is made of metals

spectrometer An instrument that separates electromagnetic radiation, such as light, into wavelengths, creating a spectrum

stratosphere The layer of the atmosphere located between the troposphere and the mesosphere, characterized by a slight temperature increase and absence of clouds

subduction Movement of one lithospheric plate beneath another

subduction zone A compressive boundary between two lithospheric plates, where one plate (usually an oceanic plate) is sliding beneath the other and plunging at an angle into the mantle

synchronous orbit radius The orbital radius at which the satellite's orbital period is equal to the rotational period of the planet; contrast with synchronous rotation

synchronous rotation When the same face of a moon is always toward its planet, caused by the period of the moon's rotation about its axis being the same as the period of the moon's orbit around its planet; most moons rotate synchronously due to tidal locking

tacholine The region in the Sun where differential rotation gives way to solid-body rotation, creating a shear zone and perhaps the body's magnetic field as well; is at the depth of about one-third of the Sun's radius

terrestrial planet A planet similar to the Earth—rocky and metallic and in the inner solar system; includes Mercury, Venus, Earth, and Mars

thermosphere The atmospheric layer between the mesosphere and the exosphere

tidal locking The tidal (gravitational) pull between two closely orbiting bodies that causes the bodies to settle into stable orbits with the same faces toward each other at all times; this final stable state is called synchronous rotation

tomography The technique of creating images of the interior of the Earth using the slightly different speeds of earthquake waves that have traveled along different paths through the Earth

tropopause The point in the atmosphere of any planet where the temperature reaches a minimum; both above and below this height, temperatures rise

troposphere The lower regions of a planetary atmosphere, where convection keeps the gas mixed, and there is a steady decrease in temperature with height above the surface

viscosity   A liquid's resistance to flowing; honey has higher viscosity than water

visual magnitude   The brightness of a celestial body as seen from Earth categorized on a numerical scale; the brightest star has magnitude −1.4 and the faintest visible star has magnitude 6; a decrease of one unit represents an increase in brightness by a factor of 2.512; system begun by Ptolemy in the second century B.C.E.; see also APPARENT MAGNITUDE

volatile   An element that moves into a liquid or gas state at relatively low temperatures; compare with refractory

x-wind   Magnetic fields that either pull material into the protostar or push it into the outer disk; these are thought to form at the inner edge of the disk where the protostar's magnetic field crosses the disk's magnetic field (also called DISK WIND)

Further Resources

Beatty, J. K., C. C. Petersen, and A. Chaikin. *The New Solar System*. Cambridge, England: Sky Publishing and Cambridge University Press, 1999. The best-known and best-regarded single reference volume on the solar system.

Booth, N. *Exploring the Solar System*. Cambridge: Cambridge University Press, 1995. Well-written and accurate volume on solar system exploration.

Comins, Neil F., and William J. Kaufmann. *Discovering the Universe*. New York: W. H. Freeman, 2008. The best-selling text for astronomy courses that use no mathematics. Presents concepts clearly and stresses the process of science.

Dickin, A. P. *Radiogenic Isotope Geology*. Cambridge: Cambridge University Press, 1995. Thorough college-level text on the uses of radiogenic isotope systems in geology and planetary science.

Dombard, A. J., and W. B. McKinnon. "Formation of Grooved Terrain on Ganymede: Extensional Instability Mediated by Cold, Superplastic Creep." *Icarus* 154 (December 2001): 321–336. Scholarly article describing a physical theory for the formation of patterns on the surface of Ganymede.

Fradin, Dennis Brindell. *The Planet Hunters: The Search for Other Worlds*. New York: Simon and Schuster, 1997. Stories of the people who through time have hunted for and found the planets.

Geissler, P. "Volcanic Activity on Io during the *Galileo* Era." *Annual Reviews of Earth and Planetary Science* 31 (May 2003): 175–211. Summary of all findings about volcanism on Io up to that date, written for a scientific audience but partly accessible to a lay audience.

Kargel, J. "Extreme Volcanism on Io: Latest Insights at the End of *Galileo*'s Era." *EOS Transactions of the American Geophysical Union* 84 (2003): 313–318. Summary for a lay audience.

Lorenz, R. D., and J. Mitton. *Lifting Titan's Veil*. Cambridge: Cambridge University Press, 2002. Richly illustrated volume describing the planning, execution, and findings of the Cassini-Huygens mission.

Norton, O. R. *The Cambridge Encyclopedia of Meteorites*. Cambridge: Cambridge University Press, 2002. Complete and well-illustrated encyclopedia of meteorites, scientifically correct and yet accessible.

Paul, N. *The Solar System*. Edison, N.J.: Chartwell Books, 2008. Begins with the origin of the universe and moves through the planets. Includes history of space flight and many color images.

Rees, Martin. *Universe*. London: DK Adult, 2005. A team of science writers and astronomers wrote this text for high school students and the general public.

Sheppard, S. S., and D. C. Jewitt. "An Abundant Population of Small Irregular Satellites around Jupiter." *Nature* 423 (May 15, 2003): 261–263. Scholarly article announcing the sudden discovery of many new small Jupiter satellites.

Sparrow, Giles. *The Planets: A Journey through the Solar System*. Waltham, Mass.: Quercus Press, 2009. Solar system discoveries told within the structure of the last 40 years of space missions.

Spence, P. *The Universe Revealed*. Cambridge: Cambridge University Press, 1998. Comprehensive textbook on the universe.

Stacey, Frank D. *Physics of the Earth*. Brisbane, Australia: Brookfield Press, 1992. Fundamental geophysics text on an upper-level undergraduate college level.

Stern, Alan, and Hal Levison. "Toward a Planet Paradigm." *Sky and Telescope* (August 2002): 42–46. Opinions from top scientists in the field about how planets should be categorized.

Stevenson, D. J. "Planetary Magnetic Fields." *Earth and Planetary Science Letters* 208 (March 15, 2003): 1–11. Comparisons and calculations about the planetary magnetic fields of many bodies in our solar system.

Thommes, E. W., M. J. Duncan, and H. F. Levison. "The Formation of Uranus and Neptune among Jupiter and Saturn." *Astronomical Journal* 123 (2002): 2,862–2,883. Scientific journal article about how these smaller planets might have managed to form so near the larger, dominating planets.

INTERNET RESOURCES

Arnett, Bill. "The 8 Planets: A Multimedia Tour of the Solar System," Available online. URL: http://nineplanets.org. Accessed September 21, 2009. An accessible overview of the history and science of the nine planets and their moons.

Baalke, Ron. "Comet Shoemaker-Levy Collision with Jupiter." Available online. URL: http://www2.jpl.nasa.gov/sl9//. Accessed September 21, 2009. Comprehensive collection of press releases, papers, data, and observations of mankind's first sighting of a giant impact.

Blue, Jennifer, and the Working Group for Planetary System Nomenclature. "Gazetteer of Planetary Nomenclature." Available online. URL: http://planetarynames.wr.usgs.gov/. Accessed September 21, 2009. Complete and official rules for naming planetary features, along with list of all named planetary features and downloadable images.

Brown, Mike. "Mike Brown, Professor of Planetary Astronomy." Available online. URL: http://web.gps.caltech.edu/~mbrown/. Accessed January 22, 2008. Brown's weekly column on planets, details on the dwarf planets including Eris, Sedna, and Quaoar, and details of his research and lectures.

Chamberlin, Alan, Don Yeomans, Jon Giorgini, Mike Keesey, and Paul Chodas. "Natural Satellite Physical Parameters." Available online. URL: http://ssd.jpl.nasa.gov/?sat_phys_par. Accessed September 21, 2009. From the Jet Propulsion Laboratory Solar System Dynamics Web Site, helpful tables and charts on the physical parameters for all known satellites of all planets.

Hamilton, Calvin J. "Historical Background of Saturn's Rings." Available online. URL: http://www.solarviews.com/eng/saturnbg.htm. Accessed September 21, 2009. Nicely illustrated timeline of ideas and discoveries about Saturn's rings.

LaVoie, Sue, Myche McAuley, and Elizabeth Duxbury Rye. "Planetary Photojournal." Available online. URL: http://photojournal.jpl.nasa.gov/index.html. Accessed September 21, 2009. Large database of public-domain images from NASA space missions.

Lunar and Planetary Institute. "Lunar and Planetary Institute." Available online. URL: http://www.lpi.usra.edu/. Accessed September 21, 2009. Wide variety of educational resources on planetary science.

O'Connor John J. and Edmund F. Robertson. "The MacTutor History of Mathematics Archive." Available online. URL: http://www-gap.dcs.st-and.ac.uk/~history/index.html. Accessed September 21, 2009. A scholarly, precise, and eminently accessible compilation of biographies and accomplishments of mathematicians and scientists through the ages.

Rowlett, Russ. "How Many? A Dictionary of Units of Measurement." Available online. URL: http://www.unc.edu/~rowlett/units. Accessed September 21, 2009. A comprehensive dictionary of units of measurement, from the metric and English systems to the most obscure usages.

Sheppard, Scott S. "The Jupiter Satellite Page," Available online. URL: http://www.dtm.ciw.edu/users/sheppard/satellites/. Accessed September 21, 2009. Complete and up-to-date orbital parameters for all known satellites of Jupiter from the Department of Terrestrial Magnetism, Carnegie Institute of Washington.

White, Maura and Allan Stilwell. "JSC Digital Image Collection." Available online. URL: images.jsc.nasa.gov/index.html. Accessed September 21, 2009. From the Johnson Space Center, a catalogue of more than 9,000 NASA press release photos from the entirety of the manned space flight program.

Williams, David. "Planetary Fact Sheets." Available online. URL: http://nssdc.gsfc.nasa.gov/planetary/planetfact.html. Accessed January 22, 2008. Detailed measurements and data on the planets, asteroids, and comets in simple tables.

Williams, David, and Dr. Ed Grayzeck. "Lunar and Planetary Science." Available online. URL: http://nssdc.gsfc.nasa.gov/planetary/planetary_home.html. Accessed September 21, 2009. NASA's deep archive and general distribution center for lunar and planetary data and images.

ORGANIZATIONS OF INTEREST

American Geophysical Union (AGU)
2000 Florida Avenue N.W.
Washington, DC 20009-1277 USA
www.agu.org.
AGU is a worldwide scientific community that advances, through unselfish cooperation in research, the understanding of Earth and space for the benefit of humanity. AGU is an individual membership society

open to those professionally engaged in or associated with the Earth and space sciences. Membership has increased steadily each year, doubling during the 1980s. Membership currently exceeds 41,000, of which about 20 percent are students. Membership in AGU entitles Members and Associates to receive Eos, AGU's weekly newspaper, and Physics Today, a magazine produced by the American Institute of Physics. In addition they are entitled to special member rates for AGU publications and meetings.

Association of Space Explorers

1150 Gemini Avenue

Houston TX 77058

http://www.space-explorers.org/

This association is expressly for people who have flown in space. They include 320 individuals from 34 nations, and their goal is to support space science and education. Their outreach activities include a speakers program, astronaut school visits, and observer status with the United Nations.

European Space Agency (ESA)

8-10 rue Mario Nikis

75738 Paris

Cedex 15

France

http://www.esa.int/esaCP/index.html.

The European Space Agency has 18 member states, and together they create a unified European space program and carry out missions in parallel and in cooperation with NASA, JAXA, and other space agencies. Its member countries are Austria, Belgium, Czech Republic, Denmark, Finland, France, Germany, Greece, Ireland, Italy, Luxembourg, the Netherlands, Norway, Portugal, Spain, Sweden, Switzerland and the United Kingdom, and Hungary, Romania, Poland, and Slovenia are cooperating partners.

International Astronomical Union (IAU)

98bis, bd Arago

FR - 75014 Paris

France

www.iau.org.

The International Astronomical Union (IAU) was founded in 1919. Its mission is to promote and safeguard the science of astronomy in all its aspects through international cooperation. Its individual members are professional astronomers all over the World, at the Ph.D. level or beyond and active in professional research and education in astronomy. However, the IAU maintains friendly relations also with organizations that include amateur astronomers in their membership. National Members are generally those with a significant level of professional astronomy. With now over 9,100 individual members and 65 National Members worldwide, the IAU plays a pivotal role in promoting and coordinating worldwide cooperation in astronomy. The IAU also serves as the internationally recognized authority for assigning designations to celestial bodies and any surface features on them.

Jet Propulsion Laboratory (JPL)
4800 Oak Grove Drive
Pasadena, California, 91109 USA
www.jpl.nasa.gov.
The Jet Propulsion Laboratory is managed by the California Institute of Technology for NASA. JPL manages many of NASA's space missions, including the Mars Rovers and Cassini, and also conducts fundamental research in planetary and space science.

The Meteoritical Society: The International Society for Meteoritics and Planetary Science
www.meteoriticalsociety.org.
The Meteoritical Society is a non-profit scholarly organization founded in 1933 to promote the study of extraterrestrial materials and their history. The membership of the society includes 950 scientists and amateur enthusiasts from over 33 countries who are interested in a wide range of planetary science. Member's interests include meteorites, cosmic dust, asteroids and comets, natural satellites, planets, impacts, and the origins of the Solar System.

National Aeronautics and Space Administration (NASA)
300 E Street S.W.
Washington DC 20002, USA
www.nasa.gov.
NASA, an agency of the United States government, manages space flight centers, research centers, and other organizations including the National Aerospace Museum. NASA scientists and engineers conduct basic research on planetary and space topics, plan and execute space missions, oversee Earth satellites and data collection, and many other space- and flight-related projects.

The Planetary Society
65 North Catalina Avenue
Pasadena CA 91106-2301, USA
http://www.planetary.org/home/
A society of lay individuals, scientists, organizations, and businesses dedicated to involving the world's public in space exploration through advocacy, projects, and exploration. The Planetary Society was founded in 1980 by Carl Saga, Bruce Murray, and Louis Friedman. They are particularly dedicated to searching for life outside of the Earth.

Index